FORECAST EARTH

WOMEN'S ADVENTURES IN SCIENCE

FORECAST EARTH

the story of climate scientist

INEZ FUNG

by Renee Skelton

Franklin Watts
A Division of Scholastic Inc.
New York • Toronto • London • Auckland • Sydney
Mexico City • New Delhi • Hong Kong
Danbury, Connecticut

Joseph Henry Press
Washington, D.C.

Author's Acknowledgments

Thank you, Inez Fung, for the many hours you spent speaking with me about your life and your work, and for generously agreeing on so many occasions to work with the book's editors and photo researchers to help us get the story right. You've led an amazing life, full of passion for your work and incredible accomplishments, and blessed by the intellectual companionship of longtime colleagues and close friends. I have been constantly impressed by your achievements, as well as the esteem and affection in which you are held by those who have known and worked with you. I have also been inspired by your tenacity and humor in the face of challenges. I am extremely grateful as well to Jim Bishop, Mark Cane, Eugenia Kalnay, Ed Sarachik, Lynda LoDestro, and Piers Sellers for taking time out of their busy schedules to speak with me about Inez Fung's life and career.—RS

Cover photo: Climate scientist Inez Fung poses against a majestic backdrop of Earth's atmosphere. Her imagination and her computer are the tools she uses to unlock the mysteries of our planet's future climate.

Cover design: Michele de la Menardiere

Library of Congress Cataloging-in-Publication Data

Skelton, Renee.
  Forecast earth : the story of climate scientist Inez Fung / Renee Skelton.
       p. cm. — (Women's adventures in science)
  Includes bibliographical references and index.
  ISBN 0-531-16777-1 (lib. bdg.) ISBN 0-309-09554-9 (trade pbk.) ISBN 0-531-16952-9 (classroom pbk.)
  1. Fung, Inez. 2. Climatologists—Biography—Juvenile literature. 3. Women scientists—Biography—Juvenile literature. I. Title. II. Series.

  QC858.F86S54 2005
  551.6'092—dc22

2005005618

Any opinions, findings, conclusions, or recommendations expressed in this volume are those of the author and do not necessarily reflect the views of the National Academy of Sciences or its affiliated institutions.

Printed in the United States of America.
1 2 3 4 5 6 7 8 9 10 R 14 13 12 11 10 09 08 07 06 05

ABOUT THE SERIES

The stories in the *Women's Adventures in Science* series are about real women and the scientific careers they pursue so passionately. Some of these women knew at a very young age that they wanted to become scientists. Others realized it much later. Some of the scientists described in this series had to overcome major personal or societal obstacles on the way to establishing their careers. Others followed a simpler and more congenial path. Despite their very different backgrounds and life stories, these remarkable women all share one important belief: the work they do is important and it can make the world a better place.

Unlike many other biography series, *Women's Adventures in Science* chronicles the lives of contemporary, working scientists. Each of the women profiled in the series participated in her book's creation by sharing important details about her life, providing personal photographs to help illustrate the story, making family, friends, and colleagues available for interviews, and explaining her scientific specialty in ways that will inform and engage young readers.

This series would not have been possible without the generous assistance of Sara Lee Schupf and the National Academy of Sciences, an individual and an organization united in the belief that the pursuit of science is crucial to our understanding of how the world works and in the recognition that women must play a central role in all areas of science. They hope that *Women's Adventures in Science* will entertain and enlighten readers with stories of intellectually curious girls who became determined and innovative scientists dedicated to the quest for new knowledge. They also hope the stories will inspire young people with talent and energy to consider similar pursuits. The challenges of a scientific career are great but the rewards can be even greater.

Contents

Climate Tracker

Inez Fung loves a mystery. The mystery in front of her right now is how Earth's climate is changing—and why.

Earth is warming up, and strange things are happening. Glaciers that have been around for thousands of years are melting. Sea levels are slowly rising. In some parts of the world, less snow is falling in winter and spring flowers are budding earlier. In others, severe storms are more common. Are all of these events connected? If so, what's happening to Earth's climate to cause them? And what other disturbing climate events lie on the horizon? These are questions Inez wants to answer.

It's hard to predict the future, but Inez's job is to try. Her tools are equations, mathematical models, and superfast computers. To understand how climate works and how it might change, she simulates Earth's natural processes—such as winds, ocean currents, precipitation, and cloud formations—inside her computer. She can change any of these variables, and then witness what could happen to climate in the next 50, 100, or even 500 years!

There's nothing about the way Earth works that doesn't fascinate Inez. She's always asking new questions and searching for new puzzles to solve. Along the way, she's helping us better understand what the future of Earth's climate might hold.

Inez Fung has done much
*of her **work** in the United States,*
as a NASA scientist.

But her journey *started halfway*
around the world in Hong Kong.

Meet Inez Fung

Inez Fung loves science fiction movies. In the summer of 2004, she rushed to the theater to see a movie that promised a look at the future. It was a future when Earth's climate had gone haywire, causing worldwide disaster.

Climate is the average weather in a place occurring over decades or even centuries. But the movie showed a lightning-quick climate change. Sea surface temperatures in the northern hemisphere dropped overnight. Ocean currents reversed direction. Huge waves crashed through the streets of New York City, sweeping away buses, cars, and people. Then a deep freeze hit. Thick layers of snow and ice covered entire buildings.

Inez watched the movie with keen interest. After all, it was a movie about future climate change—her science specialty. But she wasn't impressed. The special effects were good, but the science was bad. First of all, climate change would never happen as fast as it did in the movie, nor in the way it was shown. Second, she says with a laugh, "When I looked at the storm system they showed on the computer screen, the parcels of air were moving in the wrong direction." Oops!

Movie producers, take note. If you're going to make a science fiction movie, get the science right—or hope that scientists like Inez Fung don't go to see it.

On an outing in Oregon, Inez Fung pauses beside a lake near the base of Mount Hood *(opposite)*. Inez has learned that everything in nature—from forests and oceans to snow and clouds *(above)*—affects climate. She is also finding that what people do can affect the climate, too.

~ Science Fact or Fiction?

Bad science or not, science fiction writers love to show a future where the Earth we know has been ruined by extreme climate change. There have been desert wastelands. One movie even showed a watery world where oceans had swallowed up the land. These views of the future are fantasy. But climate change is not.

Earth's climate has shifted many times during the planet's 5-billion-year history. During the ice ages, thick sheets of ice covered large parts of Earth's land surface. But ice ages were part of Earth's natural climate cycles—they happen periodically. The Earth cools. Ice sheets form and spread. Then the ice age ends and Earth has a warmer period, like the one we're enjoying now.

But is something throwing off these natural cycles? Scientists see evidence of climate extremes that have never been seen before. The problem now, however, is the opposite of a deep freeze. The 20th century was the warmest in the past 1,000 years. And 9 of the 10 warmest years in the last 150 have occurred since 1990. Droughts (long dry spells) are more severe in some areas and snowfall has declined in others, which is a kind of drought in winter. Some trees are budding earlier in the spring. And several plant species have even started to grow in places that were once too cold for them. Climate scientists are looking at this data and asking, "Why?" Will natural cycles bring the deep freeze of an ice age back again? Or have human actions changed the cycles of climate forever?

Carbon dioxide is produced naturally, but human activitiy has greatly increased the amount of this gas in Earth's atmosphere. Vehicles alone produce almost one-fourth of the carbon dioxide released into the air.

Inez Fung is one of the scientists asking the questions. She is a climate scientist. Her job is to study what determines climate, as well as how and why climate changes over long periods of time.

The data that scientists are seeing points to people as the main culprits behind the climate change. We drive cars and trucks. We use electricity to run everything from computers to air conditioners. Our factories turn out the many products we use every day. All of these activities require the burning of fossil fuels. These are fuels such as coal, oil, and natural gas. They formed from living organisms that died millions of years ago, so they all contain carbon. As fossil fuels burn, they give off carbon dioxide as a by-product—and that's the problem.

Carbon dioxide is a greenhouse gas—a gas that helps trap heat in Earth's atmosphere. (*See box, pages 4-5.*) Since the widespread use of the coal-burning steam engine began in the early 1800s, we've been burning an increasing amount of fossil fuel. Carbon dioxide has continued to build up in the air, and so have other greenhouse gases. The concentration of carbon dioxide in Earth's atmosphere is

Sunlight reacts with air pollution to create a cloud of smog around cities such as Los Angeles. Some compounds in smog, such as ozone, are greenhouse gases.

now the highest it has been in more than 400,000 years. The added carbon dioxide has also made Earth's average temperature climb. During the past century, it has risen an average of about 1°F. Warming in some areas is much higher.

Climate scientists such as Inez are already seeing the effects of this change. Arctic sea ice has thinned. Mountain glaciers have

What Is the Greenhouse Effect?

Earth would be a frozen wasteland if not for the greenhouse effect—the process that keeps Earth's lower atmosphere warm enough to make life on the planet possible. Without it, Earth's average surface temperature would be about 54°F colder than it is now.

You can often feel the warmth of the Sun on your skin. But direct heat from the Sun

A portion of the Sun's radiation (yellow arrow) passes through the atmosphere and reaches Earth. Some of that radiation is absorbed, heating Earth's surface. Infrared radiation (red arrows), or heat, is emitted back into the atmosphere. That energy causes molecules of greenhouse gases to vibrate and produce heat of their own, warming the lower atmosphere and Earth's surface.

alone is not what keeps Earth warm. In addition, heat from Earth's surface gets trapped and held in place by the atmosphere. A similar process is at work when a sleeping bag traps your body heat and keeps you warm on a frosty night.

So how does the greenhouse effect work? To understand, you have to know about electromagnetic radiation—the form in which solar energy reaches Earth. This radiation streaks through space in invisible waves that move at the speed of light. The distance from the crest of one wave to the crest of the next is called the *wavelength* of the radiation. Wavelengths range from millionths of an inch for very high energy gamma rays to several miles for the longest, lowest energy radio waves.

The Sun emits mostly high-energy shortwave radiation. This includes the light we see, or visible light. It also includes ultraviolet radiation—the invisible energy that can give you a sunburn even on a cloudy day.

Few substances in Earth's upper atmosphere have the molecular

shrunk. Four fifths of the glaciers in Montana's Glacier National Park have disappeared in the past 100 years. The loss of land ice and the warming of oceans have caused a slow but steady rise in sea levels. The number of days with frost has also decreased over most areas of the Earth.

Inez and her fellow climate scientists are doing research to

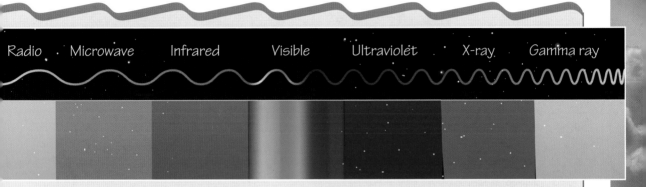

| Radio | Microwave | Infrared | Visible | Ultraviolet | X-ray | Gamma ray |

structure to absorb shortwave radiation. So about half of this radiation passes right through and strikes the planet's surface. Earth's surface does absorb shortwave radiation, but then emits it at a different wavelength. Because Earth's surface temperature is much lower than that of the Sun—only about 60°F compared to the Sun's 11,000°F—the surface emits mostly lower-energy longwave radiation, or infrared radiation, into the atmosphere.

Greenhouse gases such as carbon dioxide and methane don't just let longwave radiation zip by them. The radiation excites the molecules of these gases, making them vibrate more rapidly and produce heat of their own. This heat warms Earth's lower atmosphere as well as the planet's surface.

Electromagnetic radiation ranges from radio waves that are miles long to gamma rays that are only millionths of an inch in length. Visible light ranges from red (longest) to violet (shortest) light. When we see all wavelengths of visible light together, as in sunlight, we see white light.

Why is this called the greenhouse effect? Because a similar principle warms a glass greenhouse (left). Shortwave solar energy easily passes through the glass panes of a greenhouse, just as it passes through Earth's atmosphere. The inside of the greenhouse absorbs this solar radiation, then emits much of it as heat. This time, however, the glass traps much of the heat inside the greenhouse, just as greenhouse gases trap heat in Earth's atmosphere.

The warming climate is affecting glaciers. In 1850 Glacier National Park had 150 glaciers. Now it has about 50. Sperry Glacier has shrunk between 1907 *(right)* and 2001 *(below)*. If warming continues, Glacier National Park could have no glaciers by 2030.

understand these changes. They have some important questions to answer. Will Earth continue to get warmer? If so, how much and how fast? What is causing the warming? How will a warmer Earth differ from the Earth we know now—and how will it affect us?

No one expects the world to turn into a desert anytime soon. But there could still be serious consequences. Climate scientists say that big changes are possible in the future based on the warming they predict now. For example, climate zones could shift, keeping some plants or animals from living where they do at present. Imagine palm trees in New York! Or the North Pole with no polar bears!

If large parts of the polar ice caps were to melt, as some scientists predict they could, sea levels could rise several feet. Low coastal areas, where many people live, could end up underwater.

Islands such as the Maldives—where the highest point is just eight feet above sea level—could disappear beneath the Indian Ocean. Important ocean currents that carry warm water from equatorial regions toward the poles, such as the Gulf Stream, could weaken, cooling some regions around the North Atlantic Ocean.

What Are Greenhouse Gases?

Greenhouse gases are the atmospheric gases that contribute most to the greenhouse effect. More than any others, these gases absorb the heat that Earth radiates. They in turn re-radiate their energy to Earth, thus warming its surface and lower atmosphere. Without greenhouse gases, Earth's heat would escape into space.

Greenhouse Gas	Natural Sources	Human Sources
Carbon Dioxide	Decay of plants, respiration in living organisms, volcanic eruptions, oceans	Burning of fossil fuels (coal, oil, gas) in electric power plants and motor vehicles, burning wood, cutting trees and other plants that remove carbon dioxide from the atmosphere
Methane	Wetlands (swamps, marshes, bogs), stomachs of cattle and termites	Rice farming, cattle raising, decay of garbage in landfills, coal mining, leaks from natural gas pipelines
Nitrous Oxide	Livestock waste, wetlands	Burning of fossil fuels in electric power plants and motor vehicles, fertilizers
Ozone	Reaction of ultraviolet radiation from the Sun with oxygen in the upper atmosphere. Release from plants and trees in the lower atmosphere.	Reaction of sunlight with air pollutants (often from motor vehicles) in lower atmosphere containing carbon and nitrogen. Major part of smog.
Chlorofluorocarbons (CFCs)	None. This is an artificial chemical made in the laboratory.	Coolant in air conditioners and refrigerators, used in fire extinguishers, in making plastic foam, and as a propellant in spray cans
Water Vapor	Evaporation from Earth's surface	

~ Foretelling the Future

Inez's job involves trying to forecast a future where climate could be much different than it is today. She studies climate patterns of the past and present. She examines the many factors that are part of the climate, including ocean surface temperatures and currents, winds, and the amount of greenhouse gases and dust in the atmosphere. Then she tries to figure out what will happen to the climate if those factors change—as well as what happens to those factors as climate shifts.

The way Inez does her research is pretty amazing. She can't use the entire planet for her experiments. There's no way to make the winds or ocean currents act a certain way so she can see how they affect the carbon dioxide in the air. Instead Inez re-creates Earth—and several of its atmospheric, oceanic, and terrestrial (land-based) systems—inside a computer as a model.

Inez calls this global model her own "curious little world." She gives that virtual world characteristics that closely match those on Earth. Then she sets the model in motion and waits to see what happens.

The idea is to understand natural climate cycles, then find out how the things that people do affect those cycles. What happens to Earth's climate, for example, as people produce more carbon dioxide or other greenhouse gases?

Why does Inez do this? "First of all," she says, "it's a giant puzzle. But it's also important. The climate changes of the past 100 years have been small; they're comparable to natural changes. During the next 100 years, though, we don't expect the changes to be that small."

Many computer models predict that the effects of a warmer Earth could be drastic in some areas. There could be many more scorching-hot days. There could be severe drought in areas that are already semidry and more drenching rain in areas that are already wet. Severe storms, such as the intense hurricanes that hit Florida one after the other in the summer and fall of 2004, could become more frequent.

So Inez tries to describe what she knows to the people who make decisions that could affect climate. She can only provide the scientific understanding and computer projections. If we're lucky, decision-makers—such as government officials and business leaders—will take note. Then they will use the knowledge to prevent some of the more damaging changes from happening.

Inez has been studying climate with computer models for more than 20 years. She has made important contributions to improving them. Her discoveries have helped scientists learn more about what influences climate, and how climate could change. Along the way she has won many awards for her research.

Inez Fung has done much of her work in the United States, as a NASA scientist. But her journey started halfway around the world in Hong Kong, where as a young girl she loved to swim in the ocean and watch clouds in the sky. Back then no one could have predicted how far she would travel or the terrific things she would do.

Growing up in Hong Kong, Inez (*above*) much preferred the blue sky and open sea of the coast to the city's crowded streets (*top*).

She spent countless hours on the **sand**,
swimming in the warm waters
of the South China Sea.

To this day, the beach and the
waters of the **ocean** remain her escape.

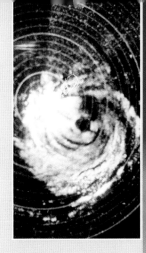

2

HONG KONG DAYS

W hen Inez thinks back to her childhood in Hong Kong, she remembers her family, her school, her friends— and typhoons (called hurricanes in the United States). Typhoons blew through Hong Kong often when Inez was a child. "I loved the fact that we got to stay home from school," she jokes now. But she also remembers typhoons as being scary storms that flattened trees and hurled solid sheets of rain past her window. Little did Inez know that weather and climate would play such key roles in her life.

Inez Fung was born in Hong Kong on April 11, 1949. She was the second of four children, with two younger brothers and an older sister.

Inez's parents were loving but traditional—"old-fashioned," she calls them. A Chinese daughter faced clear boundaries, she says: "Go to school. Study. No wild behavior. You just didn't get into trouble."

Inez's earliest childhood memories are of the beach house where her family lived on the southern shore of Hong Kong Island. She spent countless hours on the sand, swimming in the warm waters of the South China Sea. To this day, the beach and the waters of the ocean remain her escape: "There's nothing like that feeling when you just submerge yourself in the sea."

Inez was raised near the ocean in Hong Kong. Among the things she remembers are powerful storms called typhoons (satellite image above) that swept over the island with howling winds.

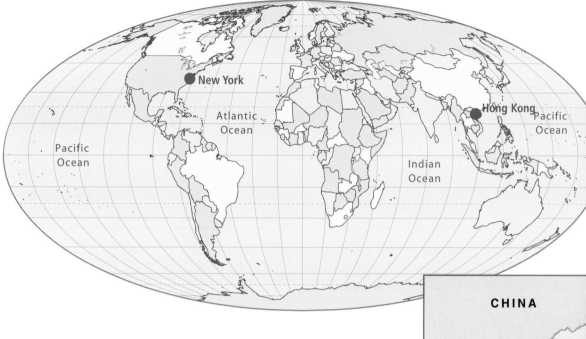

The map above shows the distance between where Inez was born (Hong Kong) and where she lived when she first moved to the U.S. (New York). Hong Kong Island, Inez's birthplace, lies off of the Chinese mainland in the South China Sea.

CHINA

New Territories

Kowloon •
•Hong Kong
Aberdeen•
Hong Kong Island

South China Sea

~ One Life, Two Worlds

Once the children were old enough to start school, Inez's parents decided to move the family to the city part of the island. The elementary school was there so it was easier for the kids to get back and forth. The Fung family moved out of their house on the beach and into a small apartment building in the city.

Inez's parents wanted to give her the best education possible. When she finished primary (elementary) school at age 11, they sent her to a private secondary school, or high school. The Fungs were not religious, but the school was a Catholic girls' school taught by American nuns. Her family overlooked the religious emphasis because of the quality of the education.

Inez remembers this as a strange time—a time when she lived in two very different worlds. "I spoke English all day at school," she recalls. "When I got home, I spoke Chinese with my family. I went to Catholic school, so I recited the rosary in English. But at home I read Buddhist mantras in Chinese for my grandmother. Tutors taught me the Chinese classics because I was in a Western school where those weren't covered."

The nuns were good teachers, but they were strict about making the Chinese students speak English—and nothing but English. "Every time they heard me say a Chinese word," Inez remembers, "it cost me 10 cents. My allowance was $10 per month. If I wanted a soda, it cost 30 cents a bottle. So unless I spoke English at school, I had no money!"

As a teenager Inez had several ideas about what she would do for a living later on. None of them had anything to do with

Inez's parents, Charles and Frances *(above)*, wanted the best education for Inez. They sent their daughter *(below, front row, far left)* to the Maryknoll Sisters School, which had high standards.

Inez began piano lessons when she was 5 years old. At age 15, she won first prize in the piano competition of the Hong Kong Schools Music Festival.

studying weather or climate. She enjoyed playing the piano and took lessons. At one point she thought she would like to become a professional pianist and play music for a living. But her teachers told her she wasn't good enough for a music career.

Inez also thought about becoming a doctor. Her parents talked to their family doctor about it. He wasn't wild about the idea. "The hours are long and the work is very hard," he said. "Being a doctor is no career for a woman."

~ Never Mind the Mosquitoes

Inez did well in school so her family expected her to go to college. To prepare her for admission to university, they sent her to a school called Kings College for her last year and a half of high school.

In Hong Kong at that time, most children didn't spend years figuring out what they wanted to do for a living. They simply did what they were good at. And by the time most students were 12 years old, Inez remembers, their teachers and parents had a pretty good idea what that was. From then on, everything and everyone steered them in that direction. For Inez the direction led toward math and science.

Kings College was an excellent school for math and science. Its teachers knew their subjects well. But the classes were tough. It was almost like attending the first year of college, not the last year of high school. Students who graduated from Kings College really knew their stuff.

For Inez the problem was this: Kings College was primarily a boys' school. Most girls in Hong Kong did not choose to study math and science in university. So there were few special schools to prepare them for college. Kings College admitted only a handful of female math and science students into the college preparatory class every year. And some boys in the school went out of their way to let the girls know they weren't welcome.

> On her first day of school, a male classmate walked up and confronted her: "You've taken the place of my friend—you've ruined his life!" he said.

Inez had no idea what to expect when she started at Kings College, but she soon found out. On her first day of school, a male classmate walked up and confronted her: "You've taken the place of my friend—you've ruined his life!" he said. "Girls shouldn't go to school. You're here only because the headmaster likes girls." *Well*, Inez thought, *nice to meet you, too!*

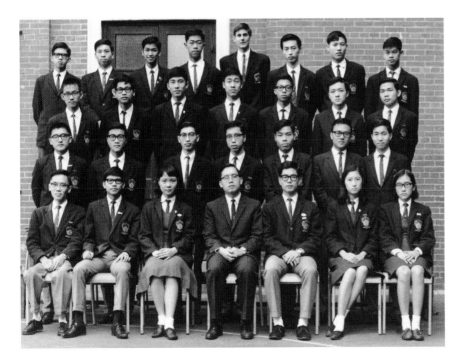

At Kings College, Inez *(first row, far right)* studied math and science. She was one of only a handful of girls in her class.

Inez went to her classes. She did well, but some of the boys kept picking on her. They played all kinds of tricks such as putting tacks on her chair before she sat down. They also made her feel uncomfortable whenever they got the chance. Some students made a point of embarrassing Inez in class. One really awful time sticks out in her memory. "In class, the students always erased the board when the teacher finished covering it with chalk," she says. "When my turn came to erase the board, the eraser had been put on a suspended light fixture. I had to climb up on the teacher's desk, reach up, get the eraser, climb back down, and erase the board—while everyone else just sat and watched."

Inez took part in a poetry recital competition when she was in high school. When this picture was taken, she had just won the top prize.

Many students would have begged for a transfer, but not Inez. Although the mean-spirited treatment hurt her, she never told her parents about it. Instead she stuck it out.

For one thing Inez felt an obligation to her family. "They thought that school would be best for me," she says. "If I didn't make it, I would bring shame to them. It would be a failure on my part and on their part."

Besides that, the competitive part of Inez refused to let her tormentors win. They would *not* see her in tears, she decided. They would *not* drive her out.

"Now that I'm older," Inez reflects, "I look back on things like that and they make me think of mosquitoes at the beach. You can slap a few and get some satisfaction. But for the most part you just have to ignore them. I was there to achieve a goal. This was just nuisance stuff along the way."

Inez also drew strength from the fact that, at the end of an awful school day, she could always return home to the love and support of her family. Inez's parents never discouraged her. They never told her she couldn't make it. "I had a very strong support system," she says. "Regardless of what happened, I never doubted my ability."

~ Facing the Future

When Inez was growing up in Hong Kong in the 1950s and 1960s, schooling dictated the course of a person's life. The smartest kids went to college.

Inez calls this type of educational system "a funnel." Once your abilities were known, everything channeled you into a certain school—and, in time, into the job you would do for the rest of your life.

It all hinged on a series of tests: Pass them and you continued in school. Fail them and you generally went to work in a job that required no college degree.

These tests were so important that teachers drilled students constantly to help them do well. One high school teacher warned Inez and her fellow students not to get sick the day of the test. Missing it, he said, would ruin their lives.

As if that wasn't enough pressure, a local newspaper printed everyone's grade on the main test. If you did poorly, Inez recalls, all of Hong Kong would find out about it. The pressure drove some students to commit suicide.

> Inez also drew strength from the fact that, at the end of an awful school day, she could always return home to the love and support of her family.

Inez wasn't worried about failing. Her parents gave her every chance to succeed. They sent her to the best schools. She had piano lessons, swimming lessons, tennis lessons, and tutors. But

As a young child, Inez lived in an area of Hong Kong Island, called Repulse Bay *(top)*. She spent much of her free time playing on the beach or swimming in the sea.

she also worked hard. Life during the school year was books, study, and more books. No hanging out with friends on school nights. No sitting around for hours watching television.

Summer was the time when Inez managed to break free. While some students took summer classes in calculus, Inez escaped to the beach. She was such a strong swimmer that her parents let her swim in the beautiful bays around Hong Kong Island. She loved it. "I had a very rigid existence—classes, school, study," Inez remembers. "But when I went to the beach, I was free. I could lie there and look at the water and the clouds. I could be in my own space."

When the time came for Inez to take her university entrance exams, she did well. She got into the city's only college, the University of Hong Kong. But at the last moment, she decided that wasn't the direction she wanted to go in life after all.

Inez had never been *away* from home.

Yet the idea of going to school
in a *foreign* land didn't faze her.

COMING TO AMERICA

3

I nez had been working toward the goal of attending the
University of Hong Kong for years. She was only 18, but
already the university saw her as a specialist in science and
math. The university would allow her to take classes such as
physics, mathematics, and chemistry. English was a different story.
A science major, the school reasoned, was not qualified to take
literature classes.

At the age of 18,
Inez left Hong Kong
to study in America
(opposite). Her
work eventually
would take her to
New York City
(above).

Inez felt disappointed—and annoyed. If the University of
Hong Kong would not let her take English, she would not let the
university take her as a student.

Did that argument alone keep Inez out of the University of
Hong Kong? We'll never know. Inez was finishing high school in
May of 1967. Around that time, huge riots broke out in Hong
Kong as followers of Chinese leader Mao Zedong confronted the
British colonial government. The rioting frightened Inez's parents.
They weren't sure whether the trouble would end or get worse.
This was no environment in which to raise children.

The political future of Hong Kong was also uncertain. The
British controlled Hong Kong at the time, but they had agreed to
return it to China in 1997. China had a communist government,
and many people in Hong Kong did not welcome life under
communist rule. The Fungs, too, worried about the future of a

21

Hong Kong under China's control. They decided to send their children to school abroad and eventually to find a new place to live and retire for themselves.

Inez had never been away from home. Yet the idea of going to school in a foreign land didn't faze her. She knew several people who had attended school in the United States. Everything had worked out fine for them. "I just thought of it as an adventure," she remembers.

Inez spent her first year in America at Utica College in central New York State. The cold and snow were a big change from tropical Hong Kong.

~ New York Interlude

By the time the Fungs decided to send Inez elsewhere for college, it was late in the school year. To go to another university that fall, Inez would have to apply right away. She and her parents rushed applications to several schools in the United States, but they

missed the deadlines. As a result, every school rejected Inez—except one. Utica College of Syracuse University, a small college in central New York State, had no deadline. The Fungs received a letter congratulating Inez on her acceptance. Inez was on her way to college in America.

In late summer of 1967, the Fungs packed up their four children and took them to North America. They brought Inez's two younger brothers to boarding schools in Canada. They stopped in San Francisco to leave Inez's older sister at San Francisco College for Women, then they took Inez to New York State. The Fungs stayed in Utica, New York, for about a day to get Inez settled at school. Then they left for home. For the first time in her life, Inez was really on her own.

Inez was in a different country, but American culture wasn't totally alien to her. "Hong Kong is a very cosmopolitan city," she explains. "I had seen American TV programs. I could get American magazines like *Seventeen*. I had gone to a school run by American nuns. So I really can't remember any culture shock."

The physical environment was another story. "The United States is vast and horizontal, with a lot of open space," she explains. That was very different from the environment of Hong Kong, a large, dense city that was built upward, not outward.

Inez was alone at school, but she wasn't all alone in New York. Distant relatives on her mother's side of the family lived in Manhattan. Inez had never met them, but when she arrived in New York their apartment became her home away from home. Whenever Utica College closed for a holiday such as Thanksgiving, she took the long bus ride to Manhattan and stayed with her aunt.

The Fungs stayed in Utica, New York, for about a day to get Inez settled at school. Then they left for home. For the first time in her life, Inez was really on her own.

While Inez was there, she became part of a large family again. She helped do household chores and babysat for her younger cousins. When she had time, she explored the lively, crowded streets of Little Italy and Chinatown. When her stay was over, her aunt put Inez on the bus back to school with a bag full of home-cooked goodies.

Inez loved the freedom of student life at Utica College. She wasn't at all homesick. "I was ready for new adventures," she remembers. But the classes were less than she had hoped for. After all those years of rigorous math and science training in Hong Kong, Inez was far ahead of the other students. "They were nice people—very kind—but the classes didn't challenge me. I got bored," she says. "I decided I wanted to transfer at the end of the year."

When Inez talked it over with her parents, they agreed. This time, there was no chance of missing deadlines. Inez applied to several excellent schools—among them Wellesley, Smith, and the Massachusetts Institute of Technology (MIT), one of the nation's top universities in science and math. Then she and her family waited eagerly to hear the news.

~ *Forks in the Road of Life*

Inez was accepted at MIT. "I got accepted because they had dormitory space," she jokes today. In those days unmarried women had to live in dormitories. Most top science schools had few places for women to live. As luck would have it, MIT had just enlarged their only women's dormitory, McCormick Hall. So, in a sense, Inez *did* get to attend MIT because of its dormitory space. Of course her excellent grades and her near-perfect score on the math portion of the SAT helped, too.

There are forks in the road of life—times when you can go one way or another. The path you pick can change the way your life turns out. Maybe there have been one or two times in your own life when you decided to do one thing instead of another, and it made a big difference. Inez's decision to apply to MIT was one of those times.

At MIT, three things helped Inez become the person she is today.

During Inez's first few years at MIT, McCormick Hall was the place she called home. It was the only women's dormitory on campus.

First, she had excellent teachers who inspired her and took an interest in her. The classes at MIT exposed her to a whole new world. Not only were her teachers the top scientists in their fields, but they were brilliant—sometimes even funny. Inez still laughs when she remembers how Alar Toomre, one of her math professors, stood in front of the class and announced they would be discussing the bending of the Milky Way. He then pulled a Milky Way candy bar from his pocket, held it up, and bent it. Professors such as Harvey Greenspan and Willem Malkus, two of her favorites, taught her much more than math or physics: They taught Inez how to *think*. They made learning exciting. "They taught the stuff in ways that related the subject to the real world," she says. "It made the subject come alive."

Although Inez couldn't have known it at the time, her favorite class—fluid dynamics, or the study of the movement of liquids and gases—would come in handy in her career as a climate scientist. It would help her understand the circulation of the

On warm spring days, Killian Court, a grassy area in the center of the MIT campus, was a great place to hold classes.

After three years as an undergraduate at MIT, Inez Fung graduated in 1971 with a bachelor of science degree in applied mathematics.

oceans and the atmosphere. "Back then I didn't know what I could apply it to," says Inez. "I just thought it was neat—and fun."

Second, Inez was surrounded by other students who were as excited about learning as she was. Some of them became close friends. Inez shared a suite with several other women. Each woman had her own small bedroom, but they all shared a common lounge. They often sat and chatted after classes. On weekends, when the college did not provide meals, they cooked and ate together in the lounge. While they dined, they talked about everything under the sun. Lifelong friendships formed.

Lynda LoDestro, today a physicist, was one of the women who cooked and shared weekend meals with Inez. They are still close friends almost 30 years later. "Inez had an energy and enthusiasm about life," Lynda remembers. "She rejected having to limit herself to one little space. She was open to whatever was there. I just loved that about her."

Inez thought about going home to stay with mom and dad, but she decided against it. She wanted to continue her science adventure in graduate school.

Third, MIT encouraged students to explore, think, and take responsibility for themselves. Back home in Hong Kong, Inez's family had controlled most of what she did. At MIT, things were different. "I went to movies a lot," she remembers. "I could do whatever I wanted—and either reap the benefits or pay the price for what I did."

Inez also explored subjects outside science and math. She took psychology and joined in discussions about how the brain recognizes objects and people. She took economics and learned about principles such as supply and demand. In electrical engineering, she was amazed by a demonstration that showed the way people's voices produce different wave patterns on a screen.

With great teachers, new friends, and room to explore, Inez enjoyed a successful three years as an undergraduate at MIT. In 1971 she graduated with a bachelor of science degree in applied mathematics. This type of math serves mainly to solve real-world problems—the design of boats and airplanes, for example, or the study of global climate change.

Now Inez faced another fork in the road. She could return home to live with her parents in Hong Kong. Or she could stay in the United States and go to graduate school. Inez thought about going home to stay with mom and dad, but she decided against it. She wanted to continue her science adventure in graduate school. There she would come to the most important fork yet in her life's road.

During the six years
Jim and Inez spent *together*

in *graduate school,*
the two became *closer* friends.

FINDING HER PATH

4

S ome people go through life with a definite goal. From the time they are young they know what they want to do—be an astronaut, design computer games, play shortstop for the New York Yankees. Inez wasn't one of those people. She was excited by the many possibilities before her. But she wasn't quite sure of what she wanted to do—at least not at first. Still, she wound up doing something pretty amazing. She became a leading climate researcher at NASA.

"Find out what you enjoy doing," Inez says today. "If you hate something, you'll never be any good at it." That was Inez's first task. She had to find out what she liked doing, not just what she was good at doing. At MIT she took a lot of different classes and exposed herself to many different subjects. She was learning which subjects she did not like. But as her last year of study neared, she lacked a plan.

At this point in her life, the adults Inez admired still influenced what she did. She tended to listen to her parents and some of the professors she respected at MIT. As graduation approached, Professor Willem Malkus talked to Inez about the possibility of getting a graduate degree in meteorology, the study of weather.

Inez made close friends during her years at MIT. One of the most important relationships was with fellow gradu- ate student Jim Bishop *(opposite)*. They shared many happy times togeth- er, including the celebration of Inez's 24th birthday in April of 1973 *(above)*.

Meteorology appealed to Inez. It related to what she was already doing, at least in part. She had taken classes in fluid dynamics, which involves the circulation of the waters of the oceans and the gases of the atmosphere. Meteorology was largely about the circulation of atmospheric gases. After a lot of thought, Inez decided to pursue a doctorate in meteorology.

All kinds of interesting people and ideas filled the MIT meteorology department. For example, a professor named Edward Lorenz had come up with something called "chaos theory." This was a description of the way some complex natural systems, such as Earth's atmosphere, behave. Lorenz didn't mean "chaotic" as in "messy" or "random." He meant it was orderly in such a complex way that it's almost impossible to predict how it will behave over a long period of time. As a result, Lorenz wondered if it would ever be possible to forecast weather more than a few days in the future.

> "I was mesmerized by the waterwheel, and by this trio of equations that explained the mysteries of the natural world."

This line of thought fascinated Inez. Lorenz had come up with three "beautiful equations," as Inez calls them. These mathematical equations described the chaos of a natural system such as the atmosphere. Professors Malkus and Louis Norberg Howard (from the math department) devised a toy machine—a waterwheel—that modeled the unpredictability, or chaos, of such a system.

One day Professor Malkus brought the waterwheel to class. It was a lazy Susan tray mounted so that it tilted upward on one side. Upright paper cups were attached all around its edges. Each cup had a small hole at the bottom to let water slowly escape. Water from a hose mounted at the top would fill the upper set of cups, and some water would drip out. As the cups on the upper side filled, the unequal weight caused the waterwheel to start rotating in one direction.

"It was the most remarkable thing to watch," Inez remembers. "The wheel rotates in one direction. Then it stops, hesitates, and rotates the other way!" It was impossible to predict exactly when the wheel would change direction or for how long.

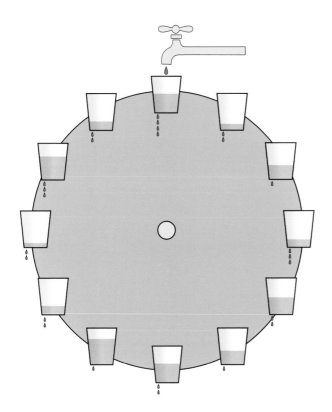

Lorenz's waterwheel demonstrates chaos theory. As water fills the top cup, the other cups slowly drain through holes in each cup's bottom. The resulting weight imbalance makes the wheel turn. As different cups fill and empty, the wheel turns one way and then another in an unpredictable way.

"I was mesmerized by the waterwheel, and by this trio of equations that explained the mysteries of the natural world. Much later, I told Professor Lorenz that I had entered the field of meteorology because of the toy illustrating his equations."

~ A New Direction

Inez's decision to study meteorology was a big step toward her present career, researching climate and the atmospheric conditions that cause climate change. But Inez had a lot of work to do. "I had no background in meteorology," she explains. "I could do the math part. But to connect that to the weather outside the window took a little time."

First, Inez says, she had to learn the language meteorologists use to talk about weather. For instance, "primitive equations" weren't equations that were crude. They turned out to be the most basic set of equations meteorologists use. Then she had to learn how to think differently about things she had already learned. In mathematics, Inez had learned that 0° meant right and 90° meant toward the top. But in meteorology, 0° is the top (north) and 90° is right (east).

In time, Inez also learned to describe weather processes in mathematical terms. "I used to think that clouds were just clouds," Inez says. "I never dreamed you could write equations to explain them—and I loved it!"

All of this appealed to the mathematician in Inez. To this day she calls certain equations "elegant" because they show complex natural processes in concise mathematical statements.

~ A Genius for a Mentor

As Inez worked through her MIT classes in meteorology, she had to choose a thesis topic. Each student in graduate school writes a book-length research paper, called a thesis, on a subject in his or her field.

Inez chose hurricanes—those monster storms she remembered from her childhood in Hong Kong. She had always noticed that rain in a hurricane falls a certain way. "If you've ever been in a

Spiral cloud bands in hurricanes *(below, left)* look like the arms of spiral-shaped galaxies *(below, right).* In her thesis, Inez changed the equations explaining the spiral arms of galaxies to explain the cloud bands of hurricanes.

hurricane, you notice it rains very heavily for perhaps 20 minutes, then the rain is lighter," she explains. "The pattern then repeats. If you're sitting in a coffee shop, you can easily wait out the heavy rain. Then you can make a dash for it when the rain lightens up."

Inez's thesis advisor, Jule Charney *(below)*, used the ENIAC supercomputer *(bottom)* to make the first successful 24-hour weather forecast.

When Inez looked at radar images of hurricanes, she saw why this happens. Cloud bands spread out from the eye (center) of these storms in spirals. In her thesis, Inez set out to explain why hurricanes are organized this way.

All graduate students get a thesis advisor. That's a professor who guides the student while he or she writes the thesis. Luckily for Inez, the department assigned her to meteorology professor Jule Charney.

You've probably never heard of Jule Charney. But you benefit from his work each time you use a weather forecast. Charney was a pioneer of computer weather forecasting. He helped prove that it is possible to use computer models to predict the weather. Charney helped develop the mathematical theory for forecasting weather in the late 1940s. In 1950 he produced the first successful 24-hour weather forecast on the ENIAC computer—the fastest computer of its day. That computer could add just 5,000 numbers per second. Today's fastest computers do trillions of calculations each second. But Charney helped lay

the foundation. Just about all weather forecasts are now done with computers.

Inez had no idea who Charney was when she first went to meet him, or that he would put her on her present career path.

What are clouds made of? How do they form? How do they move through the atmosphere?

The meteorology department was on the 14th floor of the Green Building, the tallest building on the MIT campus. Charney's office was at the end of a corridor. It had a great view of Boston across the Charles River. "I went to his office and announced, 'I'm your new student,'" Inez remembers. "He seemed like an okay person, but at that time I had no idea what he had done." To Inez, Charney was just another professor.

Charney was absent quite a bit when Inez began her classes. He traveled much of the time, doing and planning research and giving talks. But she started to work closely with him when she chose a thesis topic. Inez found Charney's style a bit strange at first. He was a brilliant scientist. He had a lot to teach students. But he made his students work for the knowledge they gained. He didn't just hand out answers. He asked tough questions that made Inez think. Then he walked away, leaving her to find the answers for herself.

Clouds can affect the global climate in opposite ways. They block incoming solar radiation and cause cooling. But they also prevent heat from escaping into space, helping warm the planet's surface. A warmer Earth could, in turn, produce thicker or more extensive cloud cover.

"When I was working on my thesis, I remember asking him what equations I had to solve to explain hurricane rain bands," Inez recalls. "He told me to approximate the vertical wind structure in the bands with two layers. Then he left on another trip." While he was gone, Inez realized that two layers would not be enough. "When he returned, I told him that what he had told me to do was wrong. He sat down next to me, smiled, and said, 'Okay, tell me about it.'"

One day Professor Charney asked Inez, "Can you imagine what it must *feel* like to be a cloud?" It sounds like a silly

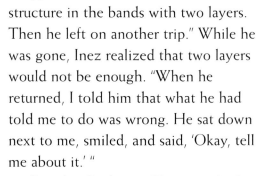

34

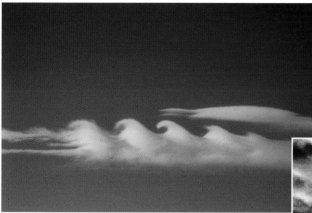

question, but it forced Inez to think more deeply about clouds. What are clouds made of? How do they form? How do they move through the atmosphere?

Without question, Jule Charney was the most important person in Inez's schooling. "Charney taught me how to think in new ways," she says. "He pushed me. He was the kind of teacher who was difficult. Sometimes, he was not very helpful—on purpose. But you learned so much because he was making you depend on yourself. He made me believe in myself—that I could do anything if I tried hard enough."

~ Friends for Life

Being Charney's student also brought Inez a whole new circle of friends. All of Charney's graduate students formed a close support group. They had lunch together each day, then they met for tea and cookies every afternoon at 3:00 P.M. At week's end, they gathered for beer on Friday afternoon. There was also a monthly get-together in Charney's library on the 14th floor of the Green Building. Everyone chipped in a dollar to buy wine and cheese. "We talked work, courses, homework," Inez remembers. "Sometimes we planned questions to ask the professors and scribbled the answers down on paper plates."

Clouds reveal lots of information if you know how to read them. Cirrus clouds *(top)* occur at high altitudes and are made up mostly of ice crystals. They usually point in the direction the air is moving. If you see altocumulus clouds *(above)* on a humid summer morning, expect thunderstorms in the afternoon.

The group was about much more than courses and school projects, though. Many of its members became close lifelong friends. Ed Sarachik had already earned his doctorate at Brandeis University. He was at MIT doing a research project under Charney's supervision. The other students all looked up to him. "Ed took care of us," recalls Inez. "We would go to him with our problems. Or he would come into our offices and ask what we were doing. I still keep in touch with him. I trust his judgment, his honesty, his friendship. When he tells me I'm going in the wrong direction, I listen first—then I argue."

Antonio Moura and Jagadish Shukla started in the meteorology department when Inez did. "We had a lot of the same classes and did our homework together every night in our offices on the 14th floor," she remembers. Both Antonio (from Brazil) and Jagadish (from India) were foreign students like herself. All three spoke fluent English. But they still got a kick out of American slang. "A herring is a fish," Inez explains, "but a 'red herring' is a false clue. We had to figure out these expressions to find out if there was any information in them that was useful."

Inez and her MIT friends Antonio Moura (*middle*) and Jagadish Shukla had lots in common. All were foreign students who enjoyed American slang—and goofing around!

Also part of the Charney group was Inez's office mate, John Willett. "He is like an older brother to me," she says.

Eugenia Kalnay became a part of Inez's inner circle. too. She was the first woman to earn a doctorate from the meteorology department at MIT. She went on to direct environmental modeling at the National Weather Service. Eugenia completed her studies before Inez entered the department. When she returned as an MIT professor near the end of Inez's graduate studies, the two women became friends.

"To be a woman at MIT, you had to be a little tough," Eugenia says today. "I admired Inez because she was smart and seemed to know where she wanted to go. She was strong and dynamic."

Popping up occasionally at the get-togethers was a Canadian chemistry student named Jim Bishop. He noticed Inez right away. "She had the greatest laugh," he says. Jim's office was below Inez's in the Green Building. She saw Jim in the halls and on the way to and from classes. At night, Inez often worked late in her office. Jim, who frequently worked late downstairs, would come up to the 14th floor to use the coffeemaker in the hall outside Charney's office.

Jim was from Vancouver, British Columbia, in Canada. Inez's parents had recently retired to that city. Inez and Jim were the only two graduate students whose parents lived in Vancouver, so they began to fly home together for holidays.

During the six years they spent together in graduate school, the two became closer friends. As they waited together at the airport to pick up or drop off Inez and Jim, the Fungs and the Bishops got to know each other, too. "Jim's parents were old Vancouver hands," Inez says. "They were extremely kind to my parents."

By the time their graduate studies neared an end in 1976, Inez and Jim had decided to get married. Their parents planned their wedding for December 30—just a couple of weeks before both of them would earn their doctorates from MIT.

Jim Bishop was working on his doctorate in marine chemistry when he and Inez became friends. They got married just before they both received their doctorates.

~ A Bright Idea

As graduation approached, Inez still had no clear idea what she wanted to do with her new meteorology degree. Then Mark Cane, a friend from the Charney group, came to her with a suggestion. Why not apply for a postdoctoral position at NASA? That was a position where she could do research under the guidance of NASA scientists.

Mark had worked at NASA's Goddard Institute for Space Studies (GISS) in Manhattan as a computer programmer before attending graduate school at MIT. Mark knew about the many opportunities at GISS. He thought it would be a good place for Inez to start her career.

Why would a scientist with a doctorate in meteorology set her sights on NASA? Most people know that NASA is the U.S. space agency, and that space exploration is a major part of its mission. But another big part of NASA's mission—the part most people know nothing about—is "to understand and protect our home planet." Hundreds of NASA scientists use data from spacecraft, satellites, and surface observation stations to study Earth's atmosphere, oceans, and land surface. A meteorology degree was a good fit.

Thirty years ago some men thought it was a waste of time to train women as doctors or scientists.

Postdoctoral positions at NASA were hard to get. Inez had to compete against hundreds of candidates from across the country. She discussed with Mark Cane what she might do. Finally, Inez decided to propose a project in which she would use satellite data to study ocean surface temperatures and currents.

~ New Marriage, New Degree

Inez kept her upcoming wedding a secret from her professors in the meteorology department. Not even Jule Charney had a clue. Thirty years ago some men thought it was a waste of time to train women as doctors or scientists. They thought women would only go off and get married, then stay home to raise their families and never use their education.

Now here was Inez about to get her doctorate in meteorology, ready to get married. She wasn't sure how her professors would react. "So I simply didn't tell them," she says.

With so much to do, Inez's friends from McCormick Hall pitched in to help her prepare for the wedding. One friend scoped out stores all over Boston, looking for a wedding dress.

Other friends chose her china patterns. Meanwhile, Inez worked on her thesis. Sometimes she felt buried beneath all the books and papers in her office in the Green Building.

Three thousand miles away in Vancouver, the Fungs and the Bishops were busy, too. They planned every detail of the wedding. As Inez puts it, "All we had to do was show up."

Inez and Jim *(left)* married on December 30, 1976, in Vancouver, British Columbia, where their parents lived. At the reception in the Fungs' living room *(below)*, the newlyweds were surrounded by family and friends.

A small group of close friends from MIT traveled to Vancouver for the wedding. The ceremony took place in the living room of Inez's parents' home. "It was a very small wedding," Inez recalls. "Jim's father had served in the Canadian military, so thanks to him we had an Army chaplain." Jim's parents threw a reception for the couple before the wedding. Then Inez's parents hosted a Chinese dinner afterward for the bride and groom and all their guests. "Every formal Chinese event has to have food," Inez says.

Inez kept her family name. "Fewer papers to fill out," she jokes. But the choice was cultural as well. In China women do not

abandon their family names when they marry. They simply add their husband's family name. Inez chose to just keep the name Fung.

With no time for a honeymoon after the wedding, Inez and Jim didn't jet off to some tropical island for sand and sun. They flew back to Cambridge. "We worked late every night in the Green Building, trying to finish our theses," she recalls. Once again, her group of friends from McCormick Hall pitched in to help. They proofread her thesis. They checked her graphs, tables, and captions. They even made sure her page numbers were correct. "I had to write thank-you notes for my wedding gifts," Inez remembers, "in-between writing paragraphs for my thesis."

Finally, Inez handed in her thesis. Then she faced her thesis defense. The "defense" is a nerve-wracking session in which the student stands before a panel of professors from his or her department. Each professor asks the student difficult questions about the thesis topic. How well a student stands up under that questioning helps determine whether or not the student passes and earns the degree.

"After my thesis defense," Inez remembers, "I was very nervous about the professors' assessment." She stationed herself in a room where she could see the elevator doors on the 14th floor of the Green Building. The professors had remained in the exam room on the 9th floor to discuss whether Inez had done a good enough job to pass. Once they made their decision, they would return here to let her know.

Inez watched the clock hands crawl. The panel was taking a long time to make up its mind—maybe too long. "In those days it was either yes or no," says Inez. "There was nothing in between. The smallest mistake was reason enough for a no."

Finally the elevator doors opened and the professors walked out, led by Charney. He smiled at Inez and said, "We couldn't decide whether to pass you or not. So we decided to give you a wedding present and let you graduate." Yes, they all knew about her marriage to Jim by now. Charney's comment was meant as a joke, but it disappointed Inez. "I had no sense of humor at the time," she explains. Reacting to her downcast expression, Charney quickly dropped the joke. The thesis was fine, he told her. The panel had enjoyed discussing it.

One by one, the professors on the panel—including Edward Lorenz—came up to Inez and congratulated her on her marriage. Professor Malkus, one of the first to suggest that she study meteorology several years before, sent Inez a note. He praised how well she had answered even his most difficult questions during her thesis defense. Inez had hit a home run. To top it all off, Inez's thesis won the Rossby Award for best meteorology thesis of the year at MIT.

With her graduate degree under her belt, a happy Inez is poised to take the next big step in her professional life.

Inez graduated from MIT with a doctorate (an Sc.D.) in meteorology in early 1977. And it was quite an unusual event. MIT's meteorology department had been established in 1928. Yet Inez was only the second woman to earn a doctorate in meteorology—second to her friend Eugenia Kalnay! She had also won the postdoc position at NASA. Her husband, Jim, won a postdoc position at Columbia University's Lamont-Doherty Geological Observatory, also in the New York City area.

With a new husband, a new degree, and a new job, everything appeared to be going well for Inez. But appearances, she soon learned, can be deceiving.

*Inez jumped at the chance
to move back to New York City.*

*She could say hello again to
her* **husband** *and her* **home.**

A NECESSARY DETOUR

5

I n January 1977 Inez and Jim packed everything they owned
and headed out from Cambridge, Massachusetts, to New York
City. They rented a small two-room apartment on West 115th
Street in West Harlem. It was within walking distance of Inez's
new job and near a shuttle bus that Jim could take across the
Hudson River to work at Lamont.

On what was supposed to be her first day of work in February,
Inez dressed excitedly. Then she made the short walk to a plain,
gray brick building at the corner of Broadway and 112th Street.
The bottom floor of the building houses "Tom's Restaurant." That's
the diner where Jerry Seinfeld and his friends hung out in the TV
series *Seinfeld*. Six floors above the restaurant are taken up by
NASA's Goddard Institute for Space Studies.

Inez was decked out in her Sunday best that day. At MIT she
had dressed in blue jeans, flannel shirts, and hiking boots. But she
was part of the working world now. So she had gone out and
bought a new suit—the first suit she'd ever worn. Because female
scientists were not the norm at GISS in early 1977, Inez surprised
the woman who met her when she arrived. "She asked me if I was
the temporary secretary," Inez remembers. "I told her I was the
new postdoc. Then I asked her where my office was."

In Manhattan,
NASA's Goddard
Institute for Space
Studies *(opposite)*
was Inez's home for
almost 15 years.
Outside of work,
New York City
offered Inez and
Jim plenty of things
to see and do such
as visiting the
Statue of Liberty
(above).

Inez was led in and introduced to her new boss. He informed her that the group she would be working with was no longer based in New York City. It had been transferred to the Goddard Space Flight Center in Greenbelt, Maryland. "He thought it would be a great opportunity for the group and for me," Inez remembers. "I just thought, *Why didn't you tell me this before now?!*"

At first, Inez was shocked and confused. Then she got angry. What was she going to do? Thinking they both had jobs in New York City, she and her husband had just moved there. Now she discovered that her job was 200 miles away.

~ Off to Maryland

Inez tried to look at the situation logically, as any scientist would. She realized she had two choices. She could accept the transfer to Maryland and live apart from her husband. Or she could turn down the job—and leave the country. Inez was not yet an American citizen. A special visa allowed her to stay in the United States because she had been promised the job (called a fellowship) at NASA. She could stay only as long as she held that fellowship. Leaving the United States wasn't a good option. Where would she go? And what about her husband? Would he have to quit his job and follow her to another country? No, Inez's best option at that point was to accept the transfer to Maryland.

Inez's decision made sense. But it began 18 months of tough times in the life of the young scientist. Inez had traded MIT's free and open school environment for the work life of a low-level government researcher. Instead of being encouraged to explore, as at MIT, she now came under the control of several layers of scientists and technicians. Inez had a task to complete: She was to study the processes that determine sea surface temperature and how it varies over ocean surfaces. This wasn't a bad topic. It interested her. It just left little leeway to explore other questions that fascinated her just as much.

The job also required strange hours. Inez needed a computer to do her work. But NASA's computer system was being upgraded.

And only three computer terminals were available. Inez—"just a lowly postdoc," as she describes herself at the time—could use the terminals only when higher-ranking scientists and programmers weren't using them. That meant she usually couldn't start working until late afternoon. She began her work each day at about 3:00 P.M., then worked late into the night.

During the 18 months that Inez worked at the Goddard Space Flight Center in Greenbelt, Maryland *(top),* she lived in the nation's capital, Washington, D.C. But she rarely had time to see the sights such as the Washington Monument *(above).*

Inez rented a basement apartment in nearby Washington, D.C., with Cathy Gebhard. Cathy was a friend who also had been Charney's assistant before moving to Washington. Inez could take the train back and forth to work because the apartment was near the end of a Metro train line. But that's where the positive part of the situation ended. Inez would pick up a carryout dinner while she worked overnight at Goddard. "My entire diet was fast food," she recalls. "There was a microwave in one of the Goddard basements—a big deal back then. Microwave popcorn was considered a special dinner!"

On Fridays Inez made her way to Union Station and boarded a train for New York City. The tracks always seemed to be under repair. The train started out, and crept along. Then it stopped. The train sat, and the passengers waited. Then the train started up again and inched along some more. Four hours often stretched to five.

When Inez finally got to New York, she was exhausted. Often she was too tired to do much more than sleep. Before she knew it, it was time to get back on the train and return to Washington.

~ Escape to New York

Inez found herself trapped in a depressing situation. The people she worked with were nice. But her job gave her no room to indulge her curiosity about things outside her assignment. There wasn't a lot of time to explore Washington, D.C. Then there were those long weekend commutes between D.C. and New York. "I thought seriously of quitting," Inez says.

Just when Inez was about to give up, Jule Charney came to her rescue. Charney had been working off and on with the Goddard group since Inez had been his student at MIT. He kept up his connection with the group, traveling to Greenbelt for meetings from time to time. During one of those visits, he gave his former star pupil some advice. "He told me my work was only part of my life," Inez remembers. "He said I should go back to New York and be with my husband."

Just as important, Charney suggested that she learn about carbon. Carbon dioxide plays an important role in warming Earth's atmosphere. In the late 1970s a small but growing number of scientists were turning their attention to the changes that

The Basics of Carbon Dioxide

Carbon dioxide is a colorless, odorless gas. In the 17th century, it was one of the first gases to be described as a substance distinct from air.

Its chemical formula, the shorthand scientists often use, is CO_2. That simply means that there is one atom of carbon bonded with two atoms of oxygen *(below)*.

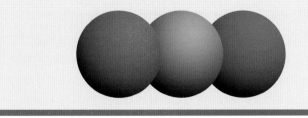

human actions seemed to be making in the amount of carbon dioxide in the atmosphere. Higher-than-ever levels of carbon dioxide—released by the burning of fossil fuels—seemed to point to a future rise in Earth's average temperature. Charney was interested in this question. Once Inez learned about carbon, Charney said they would work on a project together.

Inez believes that Charney backed up his advice with quick action. She still doesn't know for sure. It could have been a lucky coincidence, but she thinks Charney talked to James Hansen, the director of NASA's GISS, about giving her a job there in New York. Inez's MIT friend Mark Cane was working on a project for Goddard at the time as well. He also talked up Inez to Hansen. Whatever the reason, Inez found herself with a transfer to GISS in early 1979. This was the very Manhattan location where she was originally supposed to work.

Inez jumped at the chance to move back to New York City. She could say hello again to her husband and her home. And she could say good-bye to those long weekend trips. What Inez could not have predicted was that her return to GISS would prove crucial to her growth as a scientist.

Inez and Jim lived on the 15th floor of an apartment building in Manhattan. This was the view from their living-room window.

Away from work, Inez and Jim
explored New York City.

They enjoyed going to movies

and taking Saturday **walks**

in Central Park.

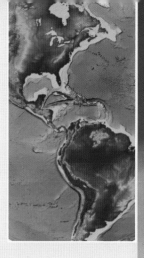

6

GROWTH AND CHANGE

For Inez Fung, the chance to move from NASA's Goddard Space Flight Center (GSFC) in Maryland to NASA's Goddard Institute for Space Studies (GISS) in New York City was a new lease on life. But the stop in Maryland was not just a short, difficult time in her life. It was an important life lesson.

Inez isn't one to sit around and complain when things don't go her way. Instead she takes a careful look at the situation. Then she finds a way to draw something positive from the experience. Sure the 18 months at GSFC had drawbacks. But it was also a crucial step in Inez's development as a scientist.

"It was a useful stop along the road because I got to study oceans there," she says today. The NASA team in New York City needed an ocean specialist. So Inez had an advantage when the director of the NASA research center in New York thought about hiring her.

Inez returned to live in New York City full-time in 1979. At work she was able to use her new knowledge of oceans and ocean currents *(above)*. During her off hours, she and Jim often took long walks in Central Park *(opposite)*.

~ A New Environment

In March 1979 Inez returned to the gray brick building at the corner of Broadway and 112th Street in Manhattan—the home of GISS. By the time she arrived, GISS had shifted its mission from the study of space to the study of Earth's environment.

A large part of GISS's original mission within NASA had been the study of atmospheric conditions on other planets, such as Venus and Mars. In the early 1970s the space agency was still sending crewed *Apollo* spacecraft to the Moon. Trips to nearby planets seemed the logical next step. GISS research would help NASA understand the climates on these planets. That would help NASA prepare to send astronauts to these other worlds.

But by the late 1970s changes were underway at NASA. The Apollo program, which had sent many astronauts to the Moon from 1969 to 1972, had ended. There were no plans to send astronauts to other planets anytime soon, so money for research to study other planets was cut.

At this point there were a number of satellites orbiting Earth. Many of them were collecting data about the planet's atmosphere. GISS needed a new focus. So its attention shifted to using data from these satellites to improve weather forecasting. GISS obtained a computer weather model from the University of California at Los Angeles (UCLA), but eventually started a new mission. The scientists at GISS decided to change the weather model to create a climate model.

By the late 1970s NASA had a number of Earth-orbiting satellites *(below)* that carried an instrument called the AVHRR, for Advanced Very High Resolution Radiometer *(inset)*. The AVHRR collected information about cloud cover and surface temperature, which GISS could then plug into climate models.

What Is a Climate Model?

You can't build one out of plastic, and you won't see one strolling down a fashion runway. This kind of model is a collection of complex mathematical equations that captures the natural processes that determine climate. Because of the huge number of equations and their complexity, they must be solved on a computer.

Climate models are some of the world's biggest computer programs, with hundreds of thousands of lines of computer code. They often run on supercomputers that can do trillions of calculations per second. Even so, it can take weeks or even months for models to produce usable results.

How do they work? A computer model divides the atmosphere and oceans into a number of stacked boxes, or cells *(right)*. Each cell represents an area hundreds of miles wide and a couple of miles high.

Scientists look at dozens of factors that affect climate in models. They include what happens to energy from the Sun and from Earth as it passes through the atmosphere, how wind circulates in the atmosphere, and the way water evaporates from Earth's surface to form clouds and precipitation. Each of these processes must be represented as a mathematical equation in the model.

Climate modelers also give each cell starting values. That is, they will give a cell a certain temperature, humidity, wind speed and direction, amount of carbon dioxide in the air, and so on. Another set of equations connects the cells by describing the way changes in one cell affect changes in others around it. Programmers then translate these equations and values into computer code.

Once the conditions in each cell are set, and the equations that describe natural processes are in place, the model is set in motion, and the days, years, and centuries advance. As the model runs—that is, as the

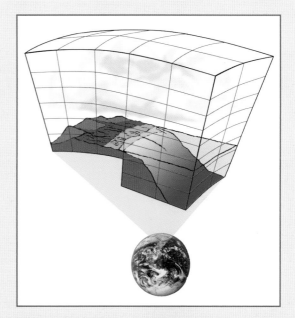

equations interact—it calculates how each cell's characteristics change over time. The model can predict climate change hundreds or even thousands of years in the future.

How do scientists know when a climate model is accurate? One way is to have it calculate past climate. Modelers check their results against the historical record. If the two data sets match, they can be more confident that their model is projecting realistic future trends.

~ "You Have to Start Somewhere"

When Inez arrived at GISS in the spring of 1979, scientists there were designing the new computer model. They hoped it would help them predict future changes in Earth's climate.

Inez knew about computer modeling because of her work in Maryland. But she had also been exposed to it at MIT. She had learned all about Jule Charney's contributions to computer modeling of the atmosphere. Using a slow, clunky computer that filled a small room, Charney and his team made the first 24-hour weather forecast in 1950. Of course, this computer was among the most advanced of the day. And predicting weather even one day ahead was a big deal back in the early 1950s. The problem was, it still took Charney and his team about 24 hours to do the calculations for the forecast. So the "forecast" came out around the same time the weather they were predicting was happening outside the window.

That forecast was not very useful for the public, but it was a great leap forward for the science of weather prediction. In the mid-1950s computers had speeded up enough to make forecasts ahead of the actual weather. And by the 1970s meteorologists could forecast weather several days in advance.

Modeling weather was one thing. Modeling climate was another. Weather models start with the conditions of the atmosphere at a certain time. Then they predict the hour-by-hour change in those conditions over a few days. Climate models show average conditions in the atmosphere over years, centuries, even thousands of years. They show the way the atmosphere responds to changes in the Sun's energy or the amount of carbon dioxide it contains. What's more, a climate model has to account for the many interconnected natural processes that affect climate. These include ocean surface temperatures, ocean currents, winds, evaporation of water from Earth's surface, precipitation, gases and dust in the air, and the size and location of clouds.

> Scientists soon came to understand that oceans have a large effect on climate.

The earliest climate models dealt only with the circulation of the atmosphere. As Inez puts it, "You have to start somewhere."

But scientists soon came to understand that oceans have a large effect on climate. Oceans cover more than 70 percent of Earth. They help distribute heat over the surface of the planet. They influence the amount of moisture in the atmosphere and they affect air temperature. They also help determine wind patterns. Because the oceans are involved in regulating climate, they had to be linked to the atmosphere in any useful climate model.

The GISS model of the atmosphere needed an ocean component. Inez had worked on ocean data at GSFC, so her boss, Jim Hansen, put her to work creating the ocean part of the GISS model. Inez quickly threw herself into this new project.

Away from work, Inez and her husband explored New York City. They enjoyed going to movies and taking Saturday walks in Central Park. They went to concerts at Lincoln Center and Carnegie Hall. On weekends they took trips to Atlantic Ocean beaches on the south shore of Long Island. At home, Inez loved to read, cook, and play classical music on her small yellow piano. Now and again she would entertain MIT pals who dropped by.

Oceans are a key part of any effort to understand Earth's climate. Inez worked on the ocean portion of the climate model at GISS so scientists could better understand ocean circulation and where ocean temperatures were warmer *(red, above)* or cooler *(blue)*.

~ Soup and Science

As Inez spent more time at GISS, she realized a lot of interesting projects were going on around her. Most of them had nothing to do with her work on oceans. Always curious and eager to learn, she took a keen interest in what the scientists around her were doing. As she got to know them, she peppered the other scientists with questions. When she found something interesting that she didn't know, she paused long enough to learn more about it. "You have to do your job," Inez stresses, "but that doesn't stop you from thinking. It doesn't stop you from asking questions. The exploration is what keeps you going."

Inez pursued her oceans assignment, but she became intrigued by the work of two women who had offices near hers. They were working with maps of vegetation and programming something into a computer. How did their work fit into the climate model?

Large forests *(below)* affect the temperature and humidity of the atmosphere over land. Data collected by satellites reveals worldwide patterns of such vegetation *(map)*.

Vegetation *does* affect climate. But climate scientists were just learning that. They had learned the value of coupling the atmosphere and the oceans. But in the early and mid-1970s, just a few lonely voices were saying that trees and other plants were important to modeling climate, too.

One such voice was that of Bob Dickinson, a meteorologist. He said it was impossible to make an accurate climate model without looking at what is happening on land. For example, Dickinson said that trees help shape the planet's climate because of transpiration. In this process, plants absorb water through their roots. Then the plants release the water as water vapor through tiny holes in their leaves called stomata. The evaporation of water from plants cools them and drains heat from the soil. This affects the temperature over large forested areas.

> In the early and mid-1970s, just a few lonely voices were saying that trees and other plants were important to modeling climate, too.

Dickinson pointed out that forests cover much of northern North America and Asia. So it would be impossible to determine the temperature of the atmosphere over Canada or Russia, for example, without knowing the amount and effect of transpiration from the huge forests in those regions.

This was not news to ecologists. But it was not widely accepted by climate scientists—at least not at first. Inez says that some scientists ridiculed the thought of including plants in climate models. "Oh, you mean include carrots and peas?" they joked. But as they came to better understand the many factors that determine climate, scientists discovered that ecosystems on land *do* influence it.

Jim Hansen knew that he needed to include land surface in the climate model. So he hired Katie Prentice, a Columbia University geography student, and Elaine Matthews, a specialist in computer mapping, to start the process. Matthews translated paper maps of the Earth's plant cover into tables of numbers. That way the planet's vegetation could be added to the mathematical model of climate. Prentice worked on a classification system that showed how vegetation would change with shifts in climate. It also showed the way changes in vegetation would modify climate.

Inez looked on until her curiosity got the better of her. "As usual, I was nosy," she remembers. "So I went and asked them what they were doing." Only about half a dozen women worked in their section of GISS and a bond formed among Katie, Elaine, and Inez. They began to eat lunch together every day, talking science over soup and sandwiches.

"At lunch, I started learning ecology by trying to understand what Katie Prentice and Elaine Matthews were doing," Inez remembers. "Day after day, I asked questions and they answered them. I learned the language of ecology. I learned their way of thinking."

Thanks to her broad knowledge of how GISS was attempting to build its computer model, Inez was able to repay the favor. She helped the two women organize and present their data in a way that would be most useful for the climate model.

~ Piecing Together the Puzzle

Inez's work with Katie and Elaine was just the first of many partnerships. In almost all of them, Inez set out to explore an area beyond the one she had been assigned. Some scientists specialize in a narrow field and never stray outside it. Inez refused to be limited by such boundaries. Yes, she was a climate scientist. But within that area she saw a lot of room for exploration.

"Some people try to put you in a box," she says. "But my mind didn't recognize the box." If a scientist is going to study carbon dioxide in the atmosphere, Inez reasoned, why not study it in the ocean, too? Why not study it on land? Why not look at the total picture?

"As long as I have an interesting puzzle in front of me," Inez explains, "I'm happy. As I went along, I would find questions or problems that interested me. Often they were questions or problems that no one else had recognized. So I pursued them wherever they went. Frequently that pursuit opened even more questions. And that's where the fun is."

Through curiosity and observation, Inez learned a lot about Earth's environment and the many processes in nature that affect climate. As she did, Inez realized that including as many of these processes as possible in climate models would make the models more complete. This, in turn, would allow the models to yield more accurate answers about future climate change.

Of course, Inez would have never realized this had she not always been open to learning. "Inez's university training was on the dynamics of the atmosphere," says MIT friend and Earth and climate scientist Mark Cane. "Then at NASA she learned a lot of chemistry and also studied the land and ecosystems. So she has a broader knowledge of the material than anyone else I can think of. She can put things together that no one else is likely to put together."

As a result, Inez played (and still plays) an important part in the development of climate models. Climate models began with the atmosphere. Then scientists added oceans. Next they recognized the influence of land processes such as transpiration. But Inez's research told her that one piece of the puzzle was still missing—Earth's biogeochemical (BGC) cycles.

> Through curiosity and observation, Inez learned a lot about Earth's environment and the many processes in nature that affect climate.

These cycles support life on Earth by moving important chemicals among the land, air, oceans, and living things. They also can affect climate. Take the carbon cycle. Because carbon dioxide affects climate, Inez reasoned that understanding how carbon moves in and out of the atmosphere was important. And including the carbon cycle—and other BGC cycles—in climate models was important.

In fact, adding these complex cycles to climate models is called Earth Systems Modeling. Today Inez Fung is known as the pioneer and main architect of this field. But it all began in the GISS offices in Manhattan with the exploration of the carbon cycle.

She worked long hours,
throwing herself into both
ocean **modeling**

and the **study** of carbon's role
in the oceans and atmosphere.

CHASING CARBON

7

In 1755 a Scottish physician and chemist named Joseph Black identified an odorless, colorless gas called carbon dioxide. Black went on to discover that carbon dioxide is one of the gases in the air, and that it is made up of carbon and oxygen. Ever since then scientists have tried to unlock the mystery of carbon dioxide's role in Earth's atmosphere.

By the early 1800s some scientists had figured out that the carbon dioxide in the atmosphere might help Earth hold in heat. This was during the early years of the Industrial Revolution. It was a time when industry was spreading in Europe and North America. Most factories burned coal as a fuel. The burning of any fuel that contains carbon—such as coal—produces carbon dioxide gas as a waste product. Until the Industrial Revolution, the amount of carbon dioxide in Earth's atmosphere had remained within certain limits for thousands of years. But the burning of coal (and then oil and gas) in factories increased as industry increased. So the carbon dioxide in Earth's atmosphere continued to rise.

In 1895 Swedish scientist Svante Arrhenius suggested that the excess carbon dioxide produced by burning coal in factories could trap more heat near Earth's surface. But no one could prove this was true.

In 1895 Svante Arrhenius *(above)* proposed that excess carbon dioxide in the atmosphere could make our planet heat up. Many decades later, Inez began to focus on the role of carbon dioxide in climate change. But she and Jim still found time to take walks on the beaches on Long Island's south shore—even during the winter *(opposite)*!

In the late 1950s, Roger Revelle *(below)* hired Charles Keeling to monitor carbon dioxide in the atmosphere from a station high atop Mauna Loa volcano in Hawaii *(bottom)*. Within a few years, data showed that the amount of carbon dioxide was rising.

By the 1950s it was clear that industry, cars, trucks, and power plants were loading the atmosphere with carbon dioxide. But what was happening to the gas? Most scientists thought the oceans were soaking up most of it. They believed the amount of carbon dioxide in the air wasn't rising that much at all. But again, no one knew for sure.

The picture became clearer in the late 1950s. That is when oceanographer Roger Revelle hired Charles Keeling to study carbon dioxide levels in the atmosphere. Keeling put a carbon dioxide monitor on top of Hawaii's Mauna Loa volcano in 1958. The volcano had thousands of miles of open ocean around it. This made it an ideal site from which to measure the average amount of carbon dioxide in the atmosphere. Within a couple of years, the Mauna Loa data showed what few had suspected. The amount of carbon dioxide was on the rise. What would this mean for Earth's climate?

Throughout the 1960s and early 1970s, a few small but determined teams of scientists were hard at work solving the carbon dioxide mystery. By the late 1970s Inez's mentor, Jule Charney, was involved with the issue. Charney headed a panel of scientists investigating the field of climate change. They asked some important questions. What research was being done on climate

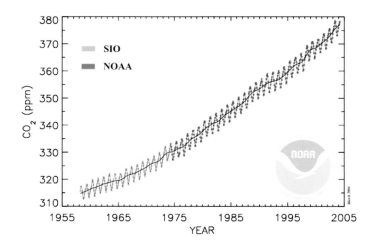

change with computer models? How reliable were these models? And should predictions of possible global warming be taken seriously?

In 1979 the National Academy of Sciences (NAS) published the panel's findings in the "Charney Report." The report stated that a rise in global temperatures of between 2.7°F and 8.0°F was likely during the next 100 years. It also urged further research on global warming because of the drastic climate change that *could* result from it.

~ On a Mission

The Charney Report came out about the time that Inez started her job at GISS. Charney passed on to Inez his interest in what was happening to carbon in the atmosphere. After all, he did tell her to study carbon. Although her new job at GISS had nothing to do with carbon, Inez took Charney's suggestion to heart. She worked long hours, throwing herself into both ocean modeling and the study of carbon's role in the oceans and atmosphere.

Inez read books and papers about carbon chemistry in the ocean. She picked the brains of her chemist friends, often at the strangest times. She went to a folk-music festival on the banks of the Hudson River north of New York City with her husband and some of their scientist friends. Carbon was on Inez's mind, as well as music. "I pestered them to explain carbonate chemistry to me," she laughs. "They were very patient."

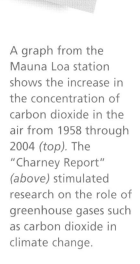

A graph from the Mauna Loa station shows the increase in the concentration of carbon dioxide in the air from 1958 through 2004 *(top)*. The "Charney Report" *(above)* stimulated research on the role of greenhouse gases such as carbon dioxide in climate change.

Eventually Inez wrote up all these notes for Charney. "I had spent a lot of time learning about ocean circulation, ocean chemistry, and the biosphere—that is, the living part of Earth's environment. I translated all that information into equations and language we would both understand."

Charney never got a chance to use the notes on carbon that Inez carefully wrote for him. Before he and Inez could work on a project together, Jule Charney died of cancer in 1981.

Inez lost an important figure in her life. Charney had set her clearly on her career path. In a tribute to him, Inez wrote: "Charney taught me to question, to formulate problems, and to focus on the finding. More important, he made me believe in myself. . . . I still write as though Charney will be reading my manuscript and will be asking questions I cannot answer until he returns from his next trip."

Charney's death was a blow to Inez, but it did not stop her. By the time Charney died, his quest had become hers. After a few years of exploring carbon, Inez was fascinated by its role in the environment, and in possible global warming. On her own she had discovered many new questions. Now she set out to answer them. As she says today, "I was obsessed."

~ To Build a Better Model

One of the things Inez does best is to think, to question, and to try to see the connections between things. "I came to learn that Earth processes are linked," she says. "Each is a piece of the puzzle. Together, they tell the whole story—I hope." These processes included Earth's biogeochemical cycles—such as the carbon cycle, the methane cycle, and so on. Inez realized that these cycles made life on Earth possible. But they also affected climate. To improve climate models, these cycles would have to be included.

Inez was among the first people to say this—and then do something about it. "I sometimes see myself as an architect," she says. "And I saw the structure I wanted to build." She wanted to build new models of these natural cycles. The models would represent the way chemicals such as carbon dioxide and methane moved among the air and living things and oceans and land. These sub-models would be added to an overall climate model. The model would then better represent the natural systems that control climate—and would be better able to describe climate change.

Inez began with the carbon cycle because it is responsible for carbon dioxide, the greenhouse gas with the biggest impact on climate. To make the model, Inez first had to know how carbon dioxide moves in and out of the atmosphere. She planned to find the sources of carbon dioxide. Then she would determine Earth's carbon dioxide sinks. These are the processes or substances that absorb carbon dioxide and remove it from the air. She would represent the sources and sinks in her carbon model. Then it could

Photosynthesis and Respiration

The amount of carbon dioxide in the atmosphere rises and falls in a seasonal cycle.

In spring and summer, carbon dioxide decreases because of *photosynthesis*—the way plants make food for their own growth. In this process, plants take carbon dioxide from the air (blue arrows), then use the Sun's energy and chlorophyll (present in green plants) to combine it with water and nutrients from the soil. This process creates simple sugars such as glucose, which plants store in their tissues as food (red and orange arrows).

In fall, leaves drop from trees and plants die. Microbes break down decaying plants, and microbial respiration releases carbon dioxide into the air (purple arrows).

be coupled to an overall climate model to better predict carbon dioxide in the atmosphere—and the evolution of Earth's climate.

By 1981 Inez had created her first crude three-dimensional global carbon model. It had an atmosphere with winds, and an ocean. Her carbon model also included the biosphere, with its forests and deserts.

By 1983 Inez had improved the carbon model she started in 1981. She knew that photosynthesis and respiration were responsible for the natural seasonal increase and decrease of carbon dioxide in the air. (*See box, page 63.*) "Each tree—whether it's an oak or a palm—fills the same role in nature," Inez explains. "In photosynthesis, for example, a tree sucks in carbon dioxide and spits out oxygen and water." To get these processes into the model, Inez wrote an equation to explain each of them. The model could then represent the movement of carbon dioxide from the atmosphere to the biosphere and back again.

What Is the Carbon Cycle?

Natural cycles make life on Earth possible. One you're familiar with is the water cycle. Earth has a limited supply of water. Still, it never runs out.

Energy from the Sun evaporates water from oceans, which rises into the air as a gas called water vapor. The vapor then cools, condenses into liquid water, and falls back to the ground as precipitation —rain or snow. Precipitation returns to oceans by rivers and as run-off from the land. The water cycle then begins again.

The element carbon is part of a cycle, too. Carbon is important because, like water, it is essential to life. In fact, carbon is called the building block of life because its compounds are a necessary part of every living organism.

Earth has a huge supply of carbon, and most of it is held in rocks. A small fraction, however, is constantly recycled, moving back and forth among Earth's living things and the planet's crust, oceans, and atmosphere. But carbon is rarely in its pure form. Instead it is found in compounds such as carbon dioxide and methane (greenhouse gases) or calcium carbonate (the material that makes up coral reefs and is abundant in rocks).

The diagram (*right*) shows how the carbon cycle moves between the land, oceans, and atmosphere.

"When ecologists look at trees, they see many different species," Inez says. "But to describe in a model how trees affect climate, species is not important. It's not what they are that's important. It's just what they do." This was a major breakthrough. Inez had created the first three-dimensional global carbon model to link the atmosphere and the biosphere, or Earth's living things.

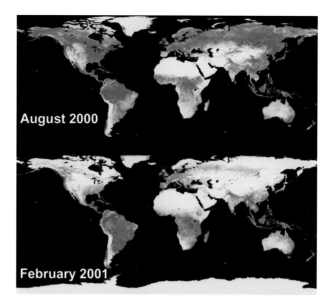

August 2000

February 2001

Satellite data shows the way photosynthesis (green) moves toward the poles in each hemisphere during summer. It moves back toward the equator in winter.

In 1987 Inez worked on a project with two other NASA scientists, Jim Tucker and her old friend Katie Prentice. Using Inez's carbon model, they applied actual satellite data on photosynthesis (a "green-ness index" developed by Tucker) and

CO_2 gas in the air.

Green plants and trees take in CO_2 during photosynthesis. They release it during respiration.

Animals eat carbon-rich plants and other animals. They breathe out CO_2. Decaying animals release carbon into the soil.

Decaying plants release carbon into the soil. Decomposers eat dead plants and animals, releasing CO_2.

Ocean water and aquatic plants absorb CO_2. Decaying aquatic plants and animals release CO_2. Wind and water break down rock, also releasing CO_2.

Over millions of years, dead plants and animals are converted into carbon-rich fossil fuels, such as coal. Burning fossil fuels releases CO_2.

the decomposition of dead plant material to show how carbon dioxide moves between plants and the atmosphere in a regular seasonal cycle. For her work on the project, Inez won NASA's Exceptional Scientific Achievement Medal.

~ Global Warming Heats Up

Inez had made carbon modeling part of her job. But she was still working on the ocean section of the GISS model. By the late 1980s the GISS model indicated a possible warming trend in Earth's future. Yet the GISS scientists could not be positive. Temperatures had dipped and risen over the previous century, making it hard to detect a clear pattern. What's more, oceans absorb heat from the atmosphere. So even after carbon dioxide levels in the air had risen high enough to cause warming, the oceans would probably delay it.

In 1981 the GISS group led by Jim Hansen had produced a paper predicting that—with the amount of carbon dioxide building up in the air—global warming should become a reality by the end of the 20th century. The study had sparked some interest outside GISS. But a bigger bombshell was to come.

The GISS findings finally exploded into the public's attention in 1988. On a June day when the thermometer topped 100°F in Washington, D.C., Jim Hansen sat down to talk to a panel of U.S. senators. He told them what he and the other scientists at GISS had been seeing over the past few years: The Earth was getting warmer. Hansen identified the culprit as greenhouse

Global warming became a headline issue in the summer of 1988 after NASA scientist Jim Hansen *(right)* told a Senate panel that global warming had already begun.

The New York Times

NEW YORK, FRIDAY, JUNE 24, 1988

Global Warming Has Begun, Expert Tells Senate

Sharp Cut in Burning of Fossil Fuels Is Urged to Battle Shift in Climate

By PHILIP SHABECOFF
Special to The New York Times

WASHINGTON, June 23 — The earth has been warmer in the first five months of this year than in any comparable period since measurements began 130 years ago, and the higher temperatures can now be attributed to a long-expected global warming trend linked to pollution, a space agency scientist reported today.

Until now, scientists have been cautious about attributing rising global temperatures of recent years to the predicted global warming caused by pollutants in the atmosphere, known as the "greenhouse effect." But today Dr. James E. Hansen of the National Aeronautics and Space Administration told a Congressional committee that it was 99 percent certain that the warming trend was not a natural variation but was caused by a buildup of carbon dioxide and other artificial gases in the atmosphere.

Global Warming: Greenhouse Effect?
Average global temperatures through the first five month of 1988. As a baseline, scientists use the global average from 1950 to 1980.

Source: James E. Hansen and Sergej Lebedeff

The New York Times/June 24, 1988

What Is Global Warming?

Over the past 100 to 200 years, the temperature of Earth's surface and the lower atmosphere has risen. This rising temperature is known as global warming. The cause? Mainly the greenhouse gases that people have been pouring into the atmosphere. *(See box, page 7.)*

During most of Earth's history, the amount of carbon dioxide in the atmosphere has gone up and down naturally. It was lowest during the cold periods we call ice ages, and highest during warm periods in between. The amount of carbon dioxide had always stayed within certain limits.

Then the Industrial Revolution began. In the late 1700s and early 1800s, countries in Europe and North America began to use power-driven machines that run on fuels such as coal. Coal contains carbon, so burning it gives off carbon dioxide. Then we began to use other carbon-containing fuels such as oil and natural gas. The amount of carbon dioxide pouring into the air increased.

Before the Industrial Revolution, the highest level of carbon dioxide in the atmosphere was about 280 parts per million (ppm). By 1958 it was 315 ppm. Today it is about 380 ppm—and rising.

Scientists have known since the 1800s that carbon dioxide had something to do with warming Earth's atmosphere. But for a long time they thought oceans would absorb any extra carbon dioxide people put into the air. Then, beginning in the late 1950s, the work of Roger Revelle and Charles Keeling showed this is not true. Revelle found that the oceans could not absorb carbon dioxide very quickly. So carbon dioxide levels would rise over time.

Keeling collected data to find out how much carbon dioxide was in the air, and how fast it was increasing. He put carbon dioxide monitors in Antarctica and on a volcano in Hawaii in 1957 and 1958. Data from the monitors soon showed that carbon dioxide was rising year by year.

By the late 1970s, climate models were predicting that temperatures would rise 0.9°F to 1.8°F by the end of the 20th century, and 2.7°F to 8.0°F by the end of the 21st century. Temperature data has backed up this prediction so far. Both Earth's surface and the upper ocean became warmer in the 20th century. In fact, it was the warmest century in 1,000 years. Several of the hottest years on record were between 1998 and 2003.

Climate models predict a warmer future, with possibly huge effects on Earth. Climate scientists anticipate more storms and rain in some areas, and more drought in others. In the future, rising oceans could flood coastal areas. Millions of people could lose their homes.

Knowing the reasons Earth's climate might change could help us prevent some of the most harmful results. For example, if rising carbon dioxide is a problem, what can people do to produce less of it? This is why the work of climate scientists is important.

gases that people were putting into the air. In the past Hansen had forecasted only that global warming *could* occur sometime in the future. This time around, he was saying much more. He was saying that it was occurring already.

"It was the first time someone claimed to have detected the warming caused by human activity in the climate record," Inez says.

This was big news. It made the front pages of newspapers and TV news shows all over the country. Global warming was now a key environmental issue.

But there was a lot of negative reaction to what Hansen had said, even from other scientists. Some scientists thought he had overstated the case. Many government officials and business leaders were not pleased. If people were causing the problem by burning fossil fuels in cars, factories, and electric power plants, people would have to make changes in the way they live to solve the problem. Pollution control devices would have to be installed on vehicle exhausts and factory smokestacks. Cars and trucks would have to switch to cleaner fuels that produced less carbon dioxide. These changes would cost money.

Industries that burn fossil fuels, such as power plants and paper mills *(below)*, pour carbon dioxide and other greenhouse gases into the atmosphere.

Hansen took a lot of heat for his findings. But his GISS team was proven right. The 1980s turned out to be the warmest decade in the 20th century up until that time. The 1990s had even warmer peaks, with several of the hottest years on record to date.

Like the rest of the GISS team, Inez supported Hansen's testimony. She also admired his boldness in going public with data that many people would find hard to accept. "I learned a lot from Hansen's courage," Inez says, "and from his insistence on doing the right thing."

~ *The Case of the Missing Carbon*

Thanks in part to the global warming controversy, research on carbon dioxide in the atmosphere soon picked up steam. In 1990 Inez was hard at work on a paper with two other scientists, Pieter Tans and Taro Takahashi. In the paper, they planned to suggest new locations for carbon dioxide monitoring stations. It all seemed rather routine. But then they plugged into the mathematical model their data on the fossil fuel sources and ocean uptake of carbon dioxide. It just didn't add up. "We overestimated the north-south carbon dioxide gradient in the atmosphere," Inez explains. In other words, the model showed too much carbon dioxide in the northern hemisphere and too little in the southern, compared to what was observed.

Most of the world's people are in the northern hemisphere. So are most of the sources of human-produced carbon dioxide. That's why more of the carbon dioxide should have appeared there. But data from actual observations showed much less than Inez's model. Where was the missing carbon dioxide?

Inez and her partners tried moving things around a little. Most scientists thought the southern hemisphere's oceans were soaking up much of the carbon dioxide that people were producing. Takahashi's data in the southern oceans backed up this idea. But when Inez put the sink there, the numbers still didn't look right. In fact the southern hemisphere sink exaggerated the problem even more.

Could the northern oceans be soaking up the excess carbon dioxide? No, that didn't work either. Scientists had loads of data on the northern oceans. It confirmed that the missing carbon dioxide wasn't there.

"We had no choice," Inez says, "but to put the sink on land in the northern hemisphere. That's exactly the place no scientist thought it could be. When our article was published in *Science* magazine, my friends called me up and said, 'You're wrong.' 'That may be,' I said, 'but you'll have to prove it to me.' And they couldn't. There was no data."

No one, in fact, could prove Inez wrong because her logic was right. Inez's mathematical model had predicted that a huge land sink in the northern hemisphere was soaking up the planet's excess carbon dioxide.

This was a landmark finding. It overthrew what almost every other climate scientist in the world had believed until then about a major part of the carbon cycle. "That paper was a turning point in the study of the carbon cycle," Inez says. "Before then, oceanographers thought the oceans were responsible for removing fossil-fuel carbon dioxide from the atmosphere. They thought that the land did very little. But here we were saying that the land and the oceans absorb approximately equal amounts of fossil-fuel carbon dioxide. It was right here, in the much-studied northern hemisphere."

> No one, in fact, could prove Inez wrong because her logic was right. Inez's mathematical model had predicted that a huge sink was soaking up the planet's excess carbon dioxide.

Critics seized on the fact that Inez's paper had not answered exactly how the carbon dioxide was disappearing. "Fingers pointed in all directions," Inez says. "The oceanographers and ecologists attacked the model. The atmospheric people argued that we could not really extrapolate from the sparse land and ocean data. The upshot is that great new observing programs went into place. Several other modeling groups followed suit. They all came to the same conclusion. Now an international project is in place. More than 15 global models are taking part."

If you want to know how carbon dioxide affects climate and people, it's important to know how it gets into and out of the atmosphere. Inez put a large piece of that puzzle in place.

"That's the amazing thing about science," Inez continues. "We didn't ask where the carbon was going at first. We were arrogant and thought we knew. But when we started down that path, what we thought we knew, we really didn't know at all.

"Every wrong turn in the road can be a detour to learning something new," Inez says. In this case, that wrong turn led to a huge discovery.

Ever since her childhood swims
in the **bays** around Hong Kong Island,
Inez had enjoyed the sea.

Now she had a house
near the **coast** of Vancouver Island.

8

A Small Detour to a Big Discovery

After carbon dioxide, Inez tackled another important greenhouse gas—methane. Like carbon dioxide, methane is a trace gas. It makes up just a tiny percentage of the gases in the atmosphere by volume. In fact there's even less of it than carbon dioxide. But it packs a huge global warming punch. Each molecule of methane has about 23 times the atmospheric warming ability of a molecule of carbon dioxide.

~ Pigs and Paddies

Before Inez could make a methane model, she had to figure out how it moves into and out of the atmosphere. The gas comes from a number of places. It's the main component of natural gas. The decay of organic matter in environments with little oxygen also releases methane. Rice fields and wetlands give off methane gas. The gas comes from the stomachs of cows, pigs, and sheep, too. People help release methane when it leaks out of gas pipelines, seeps out of coal mines, or rises from landfills as garbage decays. Methane stays in the atmosphere several years before it breaks down there, or in soil.

In their new home in Victoria, Inez and Jim were only a 5-minute walk to the Pacific Ocean (opposite), where they loved to kayak—at least on days when the ocean was calmer. They also traveled throughout the western United States, visiting places such as Arizona's Canyon de Chelly (above).

For Inez the first step was to map the location of major sources and sinks. "We mapped wetlands. We mapped cows. We mapped pigs and camels. We mapped landfills. We mapped rice fields. You name it. If it produces methane, we mapped it," Inez says. But the mapping was tedious and, as Inez puts it, "not as cool as exploration."

Methane comes from many sources, including cows *(right)* and rice fields, or paddies *(below)*. Bacteria in the stomachs of animals such as cows, buffalo, and camels are powerful methane factories. So are microbes in the waterlogged soil of rice paddies.

The data compilation may not have been cool, but it was an important job well done. Inez published a three-dimensional global methane model in 1991. "I think it is one of the most comprehensive models published," Inez says. "There are variations on my methane model, but people are still using the data my group produced."

To see how changes in methane affect climate, the next step would be to connect the methane model to a climate model. So far that hasn't happened. Inez says this is because, unlike carbon dioxide, it is hard to predict how methane sources will change over time. "I do not know how to write an equation that summarizes the way the population of cows and sheep will change," she explains. "It's also hard to predict the leakage from natural gas pipelines or to forecast how rice fields will be irrigated or fertilized. These things affect the release

of methane into the atmosphere." Someday someone may be able to do the work needed to add the methane model that Inez created. But for now, that research remains undone.

~ Good-bye, New York

In the early 1990s Inez was busy finishing old projects and proposing new ones at GISS. By 1993 she had lived on the East Coast of the United States for 25 years— longer than she had lived in her native Hong Kong. Inez was now an American, having become a U.S. citizen a few years earlier, in 1986. But she was thinking of leaving the United States for a chance to be near her family again.

Inez's parents had moved from Hong Kong to Vancouver, Canada, when she was in graduate school. They were in North America but still 3,000 miles away. As a result, Inez did not get to see them very often.

Then a job offer came out of nowhere. By the mid-1990s, Inez was well-known and respected as a climate scientist. She was the kind of person who universities like to hire for their teaching staffs. The University of Victoria offered both Inez and her oceanographer husband teaching jobs. Victoria, a city on the southern tip of Vancouver Island, was only a 90-minute ferry ride from both sets of parents in the city of Vancouver. The offer was worth looking into!

Inez wasn't sure what to do. She enjoyed her work at NASA. But other forces were pulling her west. "I left home when I was 18," Inez says. "My husband left when he was 22. The offer of two full professorships near both of our homes was a chance to be with our parents while we still had the time."

When Inez and Jim left New York, Inez's colleagues at GISS threw a going-away party for them.

For Inez and Jim, saying farewell to New York meant leaving behind sparkling views of the city. But Inez was able to remain connected to her research projects.

Jim Hansen didn't want Inez to go. He told her she would do her best science work if she stayed at NASA. Even when Inez insisted on giving Victoria a try, Hansen did not let go completely. "He pointed out to me that I could make a greater scientific contribution if I remained a full-time researcher," Inez says. "When I told him I had decided to leave after all, Hansen was very supportive. He gave me an extended 'leave of absence without pay' from NASA." That way, Inez could remain connected to her research projects.

Although she chose to leave GISS, Inez appreciates the room she had there to grow as a scientist. "Hansen never told me what to do, or what not to do," she says. "It wasn't rigid. So there was room for exploration. There were a lot of very smart people doing interesting research. Even though I didn't understand everything at first, I slowly learned a lot of things.

Three thousand miles to the west, Inez and Jim settled into a ranch house in a suburb of Victoria. Life was much quieter and slower than it had been in Manhattan.

Several generous colleagues were willing to share what they knew with me. Most important, Hansen was the kind of boss who recognized important science even when he didn't understand all the details of the work."

Some scientists have a mentor who guides them throughout their career. Inez had lost hers with the death of Jule Charney in 1981. But she made up for it with her natural curiosity, hard work, and determination. "I didn't feel as though I needed an official mentor or role model," she says. "The learning and exploration were exciting—and eventually they led somewhere. I was also lucky that my friends were brutally honest and trusted me with what they thought."

~ Off to Canada

Inez and Jim were excited about their new teaching positions at the University of Victoria. But leaving New York City was difficult. After 15 years there, Inez had become a real New Yorker. "My neighborhood knew me," she remembers fondly. "The chocolate lady, the butcher, the deli man, the dry cleaner—even the panhandlers." She had to give up cappuccino and cannoli on the Lower East Side, Broadway plays, and popping into shops for late-night takeout on the way home from work. She also gave up her comfortable apartment in Manhattan.

Three thousand miles to the west, Inez and Jim settled into a ranch house in a suburb of Victoria. Life was much quieter and slower than it had been in Manhattan. That took some getting used to. The house had a pool in the backyard, where Inez could swim—when the local wildlife wasn't using it, that is. "One time a family of raccoons had taken quince from the tree and were washing them in the pool," Inez remembers. "Another time I found a river otter sunning itself by the pool. It was large and a bit scary. But it looked so comfortable we didn't want to disturb it. Then there were the ducks that set up shop in the pool from time to time."

Ever since her childhood swims in the bays around Hong Kong Island, Inez had enjoyed the sea. Now she had a house near the

One nice thing about living on Vancouver Island was how easy it was to go kayaking. Here Jim floats offshore while Inez snaps his picture.

To celebrate her father's birthday in 1993, Inez and her family traveled to Alaska's Glacier Bay.

coast of Vancouver Island. When they had free time, Inez and Jim would each grab one end of their two-person kayak and carry it down to the water. "We could walk it five minutes from our house and go for an easy paddle along the coastline of Victoria," Inez recalls. "As we glided along, seals popped their heads up to look at us."

Inez enjoyed the natural environment of Vancouver Island, with its crystal-clear inlets and dark conifer forests. "Sometimes we'd go walking on the beaches on the south coast of the island," she says. "There were beautiful bays, huge logs on the beach, and forests—misty and mysterious. The sea life was rich. Nature was very much alive there."

~ From the Classroom to the Edges of the World

Inez had never taught school before coming to Victoria. But she loved the idea of passing on her enthusiasm for science to students. Inez taught an introductory Earth science course, among others. Some dedicated and enthusiastic students passed through her classroom. But she missed working closely each day with a team of people who were driven to find the answers to important questions.

Inez's link with NASA was her research lifeline. She kept in touch with her research team of postdocs in New York. She even invited them to Victoria from time to time to finish papers under her guidance. "Each postdoc knew that he or she could come to Victoria for two weeks," Inez explains. "They would come prepared, with their graphs and tables organized and their paper outline ready. They knew that all I would do while they were here—except for my teaching obligations at the university—was work on the paper with them."

Inez loved these times. There was spirited discussion and debate about the science. It was also a chance to visit. Whenever one of her guests needed a break, he or she could take off in Inez's little red Honda Civic for a drive to the west shore of the island. There they could walk in the rain forest or watch for whales in the Pacific Ocean. "Victoria is a far cry from Manhattan," says Inez. "Most of them thought it was the edge of the world."

Yet important science also got done. In 1995 Inez teamed up with postdoc Ina Tegen to develop a dust model. Yes, even lowly dust is a vital part of the climate change puzzle. Wind, especially in dry areas, picks up dust and can loft it high into the air. Fine dust particles can stay suspended there for long periods and travel long distances around the globe. These particles can reflect solar energy back into space, cooling the Earth's atmosphere.

Inez's 1995 paper on the subject kicked up a little dust storm of its own. Inez and Ina stated that as much as half of the dust in the atmosphere might come from human activities that destroy plants and disturb the soil. These activities include plowing farm fields, letting cattle overgraze grasslands, and cutting forests.

In 1995 Inez wrote that people might be responsible for as much as half of the dust in the atmosphere. Cutting forests *(above)* and plowing fields *(left),* for example, both remove vegetation and disturb the soil.

What could this mean for future climate change? Unusually high levels of dust clouded the atmosphere during the last ice age. Could more dust mean another ice age sometime soon? Not likely, especially with the high amounts of carbon dioxide and other greenhouse gases in the atmosphere. But Inez's dust model is now linked to the NASA GISS climate model. It may one day help identify the link between dust and changing climate.

~ A Tempting Offer

After about three years in Victoria, Inez received a new teaching opportunity. The University of California at Berkeley (UC Berkeley) was creating a Center for Atmospheric Studies. It needed a director. Would Inez like to apply for the job?

Her first answer was no. *I don't do director stuff,* she thought. *Too many meetings. Too much paperwork.* Several of Inez's friends urged her to take the position. UC Berkeley was one of the most respected universities in the United States. In spite of this encouragement, she just wasn't interested.

Then a comment from a friend changed her mind. "He knew the right button to push," Inez says. "He said, 'You have to apply because you have to prove that a woman can do this. It's not just an old-boy thing.'" That got Inez thinking: *Maybe I should consider it.*

Inez decided to give a seminar at Berkeley on her recent work on the dust cycle. *What better opportunity,* she thought, *to take a close look at the university and learn more about the position?* "When I got there," she recalls, "I said, 'Wow—I can work with these people!' I got totally excited. Now I *wanted* the job."

After chatting with some faculty members and students, Inez realized that UC Berkeley offered wonderful opportunities for exploration and learning. "People at Berkeley were interested in teaching me what they knew," she remembers. "They were also interested in learning new topics. After my talk on the dust cycle, someone asked if I could relate my findings to what Charles Darwin had found on the voyage of the *Beagle.* Geologist Walter Alvarez showed me rocks with layers of fallout from the meteor

impact that killed the dinosaurs. The faculty and students were up-to-date outside their own research areas. There was a nice kind of energy. The students were smart and I thought I could learn new things."

Inez applied for the position and won it. In 1997 she became the first director of UC Berkeley's Center for Atmospheric Studies. Her husband, Jim, landed a job in oceanography research at Lawrence-Berkeley National Laboratory nearby. Once again Inez and Jim packed up their belongings and moved, continuing their lifelong quest for more challenging opportunities in science.

Before moving to California to take her new position at UC Berkeley, Inez returned to Hong Kong to visit family and friends.

Inez doesn't fear criticism or occasional setbacks.
"They're not fatal," she says.

Each one *teaches* you *something*.

9

ICE AND CARBON

For Inez a place at the end of the Earth might also hold a piece of the climate puzzle. The Vostok Research Station sits on the Antarctic ice sheet, in the spot where Earth's coldest temperature, -129°F, was recorded. Working from 1992 to 1998, scientists removed a two-mile-deep ice core from the ground at Vostok. The core holds layers of gas bubbles containing air from Earth's ancient atmosphere. The core also contains snow, dust, and chemicals laid down over 420,000 years of the planet's history—including its last four ice ages.

Climate models need new phenomena to explain, Inez says. The Vostok core holds challenges for climate modelers. For starters, it reveals the concentration of greenhouse gases in Earth's atmosphere during past ice ages and the warmer periods between. Can models use these clues to determine past climate?

~ Answers in the Ice?

That ice core sample may also hold clues to Earth's future climate. "It's what I think about when I'm just walking around town," Inez says playfully.

At a conference in Australia in 1997, Inez tagged along with some ecologists doing fieldwork at Daine Tree River *(opposite)*. About the same time, scientists at the Vostok Research Station *(above)* were removing part of a 2-mile-deep ice core that contained gas bubbles of air from Earth's ancient atmosphere.

For Inez the Vostok ice core is a deep mystery just begging to be solved. The core sample shows that carbon dioxide in the atmosphere sank to about 200 parts per million (ppm) at its lowest and rose to a high of 280 ppm during the years represented by the core. Today the concentration of carbon dioxide in the air is about 380 ppm. That means it is higher than it has been in almost half a million years.

"We've already exceeded the ceiling in the ice core," Inez says. "So we'd like to know how resilient Earth's climate is." In other words, how much more carbon dioxide can we pump into the air before catastrophe results?

Inez has lots of questions that could stimulate new research. For example, the core could answer new questions involving the role of dust in climate change. It shows high dust levels during the ice ages. That finding seems to make sense. Increased dust would scatter solar radiation rather than letting the ground absorb it. That would add to the cooling of the Earth's surface. But could there have been another cooling mechanism —one that had to do with the iron in dust?

As scientists know, the dust particles that settle on ocean waters contain iron. Iron helps plants in the ocean grow better. In a 2000 paper, Inez showed that the iron stimulates the growth of sea plants because it helps them use nitrogen better. Plants need nitrogen for photosynthesis. The increased photosynthesis in sea plants makes them use more carbon dioxide. The level of carbon dioxide in surface water drops. So the plants start to suck it out of the atmosphere. The possible result? Earth is left with less ability to hold in heat. So increased dust in the oceans could lower carbon

The Vostok Research Station in Antarctica has operated year-round since the late 1950s. The tall drilling tower is used to bring up samples of the Antarctic ice sheet.

84

dioxide in the atmosphere, thus cooling Earth's surface.

As Inez had noted in her 1995 paper on dust, people may be raising the amount of dust in the atmosphere through actions that strip vegetation from the surface and expose bare soil. What's more, as Earth's climate warms, dry areas could expand. Winds can pick up dry, exposed soil in these areas and scatter it into the atmosphere.

At the National Ice Core Laboratory in Denver *(above)*, ice cores from Vostok *(inset)* and elsewhere hold secrets to half a million years of Earth's climate.

Global warming leading to global cooling? Interesting question. But Inez doesn't think it's likely—at least not now. "There is no way to get back into an ice age," Inez believes, "even if we're due for one. That's because our feet are on the pedal and we're loading the atmosphere with an increasing amount of carbon dioxide." There have been sudden shifts in climate in the past. Climate scientists aren't sure what triggered them. But in Inez's view, people are testing the limits of what the atmosphere can bear before a sharp climate shift occurs again.

What shape those shifts will take fascinates Inez. "Around the time of the last ice age," she says, "we know the amount of carbon dioxide in the atmosphere went down and the climate cooled. We also know there was more dust. Perhaps more iron in the ocean from the dust stimulated marine productivity. This pulled more carbon dioxide out of the atmosphere, and it cooled off some more."

Of course with Inez, this just leads to another question. "Why didn't the cooling just go on forever?" In other words, what put the brakes on the cooling that caused the ice age, allowing Earth to warm up again? Inez doesn't know. But as she walks around town, she is forming questions in her mind that one day might lead her to an answer.

> Many scientists thought her theory was absurd—until new data proved her right.

~ Just One More Question

Inez likes to formulate theories, then put them out for debate. She has often been proven right in the end. But that doesn't mean she hasn't faced her share of criticism. Remember the 1990 paper in which she said that the land in the northern hemisphere was absorbing the carbon dioxide that seemed to be "disappearing" from the atmosphere? Many scientists thought her theory was absurd—until new data proved her right.

Then there was the more recent paper in which Inez suggested that human activities might be causing up to half the dust in the atmosphere. "Because of that hypothesis, I might as well have painted a bull's-eye on my body," Inez says, laughing. "It was like target practice. But because of that, we now have much better observations of dust."

Inez doesn't fear criticism or occasional setbacks. "They're not fatal," she says. Each one teaches you something. In fact she enjoys a little controversy now and then. "If I give a talk and everyone just sits there and agrees with me, that's boring," she says. "I enjoy the give-and-take. It's like a game of tennis against a worthy opponent."

The scientist also has a subtle sense of humor and can give as well as she gets. "I was at a meeting, and a young man was giving a talk about iron input to the ocean," Inez recalls. "He was making elaborate claims based on a dust map he had. So I stood up and asked, 'How much confidence do you have in that map?' And he said, 'I don't know. Some woman did it.' Then I said, 'How much faith do you have that the *woman* did the right thing?' At that point, everyone in the meeting started laughing. They knew I was the woman responsible for the map."

~ Back to Carbon

Inez also has a reputation for speaking up and fighting for what she believes—a far cry from the quiet young woman who came to America from Hong Kong 40 years ago. The United States and some European countries are now running a new type of carbon model because Inez pushed the idea a few years ago.

Now that Inez is no longer at NASA, she doesn't work with the GISS model. Like many American university scientists, she works with the Community Climate System Model (CCSM) at

During a 1997 conference on carbon dioxide in Australia, Inez and other scientists explored a mangrove forest. Mangroves form an important coastal habitat for fish and other creatures and also play a role in the carbon cycle.

the National Center for Atmospheric Research (NCAR). A few years ago Inez and fellow scientist, Scott Doney, convinced NCAR scientists to add a carbon model to their climate model. "I told them, 'You really need this. You just don't know it yet.'"

Like her older model, Inez's new model depends on the operation of the carbon cycle. The difference is, in the old model she had to specify a value for atmospheric carbon dioxide. Then the model would generate a climate scenario. This new model predicts the carbon dioxide. "Who knows," she jokes, "It could go wrong in new and interesting ways."

This new model builds on the work that Inez and several of her colleagues have done over the past 20 years. NASA scientist Jim

Inez is pictured with her colleagues Piers Sellers *(left)* and Bob Dickinson. These three scientists, along with Jim Tucker, developed the blueprint for studying how the biosphere works on a global scale.

Tucker figured out how plant leaves use energy during photosynthesis. He developed a way to use satellites to track how photosynthesis changes on Earth's surface from month to month and year to year. *(See map, page 65)*. When he was still a NASA scientist, Piers Sellers figured out how temperature, sunlight, and humidity affect the way plants absorb carbon dioxide from the air as they grow and how they cool by losing water. Piers then helped explain Tucker's map by writing equations that explain the amount of photosynthesis that occurs on any part of Earth's surface at any time. Bob Dickinson determined the way plants use water and how this water affects climate. Inez then figured out how all these processes affect the amount of carbon dioxide moving into and out of the atmosphere. For the first time, these scientists were able to explain mathematically how the biosphere works on a global scale. Inez could use these equations in the development of her new climate-carbon model.

The new model is unique because it takes into account the fact that climate helps determine the amount of carbon dioxide in the air—just as the amount of carbon dioxide in the air helps determine climate. "Changes in temperature and rainfall change how fast plants undergo photosynthesis and how fast plants decompose," Inez explains. That affects the amount of carbon dioxide being sent into or sucked out of the air. "Global warming would also slow down the ocean's circulation, and how much carbon dioxide it removes from the air. These factors would change not only the level of carbon dioxide in the atmosphere, but climate itself."

> "I'm getting deeper into the *why* question now," Inez explains. "*Why* are carbon dioxide levels changing?"

"Inez was the first to propose this," says Ed Sarachik, a climate scientist and friend from Inez's MIT days. "You don't just couple the atmosphere to the ocean and land. You also couple the carbon cycle to the atmosphere, the ocean, and the land. So you get the concentration of carbon dioxide as the climate changes."

It's all a matter of seeing more and more of the total picture. "I'm getting deeper into the *why* question now," Inez explains. "*Why* are carbon dioxide levels changing? To answer that, I need to know how much fossil fuel we are burning each year and how the land system and the oceans pick it up. I looked at dust because that changes how sunlight heats both the ground and the atmosphere. Dust delivers iron to the ocean, and that changes the biology of the seas. That in turn changes the carbon dioxide abundance in the atmosphere, and so on. So I'm steadily embracing more and more of the problem."

This is one of Inez's strengths. She has knowledge in several areas and can put together many of the pieces of the climate puzzle—pushing forward our knowledge of climate change.

Inez has been **honored**
several times in her career

for pioneering **research**
in the field of global climate change.

MOVING AHEAD

10

From Hong Kong to Boston to Washington to New York to Victoria—and now to Berkeley. Some people never stray far from the place where they were born. But Inez Fung has crisscrossed oceans and continents to pursue her education and career. Now in her late 50s, Inez has settled into a comfortable old house in northern California's Berkeley Hills—"about 20 minutes puffing distance uphill from campus," as she puts it.

Inez and her husband, Jim, both have busy careers. As an oceanographer, Jim is working to understand the carbon cycle in the ocean. "If you want to distinguish what I do from what Inez does, she takes care of everything up to the ocean's surface," he says. "I'm trying to unravel what goes on below." Jim often spends weeks at a time at sea collecting data on the ocean from ships.

When Jim and Inez are both at home, they love to take walks in Berkeley or make the 90-minute drive to the Pacific coast for strolls along the beach. They also travel when they can. On a recent trip to Hawaii, Inez—always the climate scientist—made a trek to Mauna Loa. It was at the top of that volcano that the first measurements of atmospheric carbon dioxide were made more than 40 years ago.

Cooking is also a passion. The kitchen in Inez's house is the site of many weekend-afternoon experiments—with food, not

In 2002 Inez was inducted into the National Academy of Sciences *(opposite)* in recognition of her pioneering work in studying the carbon cycle and other factors involved in climate change. Now in her late 50s, she is still looking for new challenges *(above)*.

trace gases. Inez and one of her best friends, Lynda LoDestro, love to get together and cook in Inez's kitchen. They choose elaborate recipes, then they try them out. It's cooking for fun. But Inez being Inez, there is always a bit of exploration, too.

The two scientist friends follow the recipes, trying to learn why certain ingredients are used—or not used. They analyze the ingredients to discover why some recipes work out and some don't. Sometimes they plan and cook dinners for large parties. "Before those, we have to practice," Inez says. At other times, the meals aren't as formal. They'll try out a recipe and invite houseguests or nearby friends and neighbors to share the meal. Sometimes not everything works out as planned. One day Inez and Lynda tried to make a lamb roast encased in a shell of coarse salt. "You pack the salt around the meat," Lynda explains. "A little moisture makes the salt start to dissolve. But then it forms a casing before it moves very far." Surprisingly, the meat isn't salty. But cooking inside the casing makes it wonderfully tender and juicy. The two friends baked the lamb, but the salt shell was like plaster. "The shell was so hard we couldn't get it off," she laughs.

> She does her best to draw students in and make the subject exciting— the way her own teachers did for her at MIT many years ago.

"I remember that," Jim says. "I'm the one who had to come in and break it open with a hammer." Luckily, they could still eat the roast. Besides, as Inez often says, failure can be a great teacher—in cooking as well as in science. "You might make a mistake, but the next time you make the dish you'll know exactly what not to do."

~ The Teacher

When she's not relaxing at home, Inez is often on the Berkeley campus in her large, cluttered office. On the wall outside her door hangs a giant satellite image of the world. Or you may find her in the classroom. Inez loves the give-and-take with her students. She especially enjoys teaching introductory courses in Earth science and environmental science. She always hopes that she can interest

at least a few students in a science career—even if that wasn't their idea when they signed up for the class. She does her best to draw students in and make the subject exciting—the way her own teachers did for her at MIT many years ago.

Inez is known for the way she stays in contact with old friends. Here she and Jim *(far left)* take her old MIT professor Edward Lorenz (with cap) and James Kirchner of UC Berkeley up to see Mt. Tamalpais near San Francisco. UC Berkeley *(below)* is now Inez's research and teaching home.

In the freshman seminar every week, Inez has lunch with the class. They talk about the atmosphere over sandwiches. "They're smart and energized," she says admiringly of her students. "A lot of them are first generation—the children of immigrants. Some of their parents are struggling. They work hard to come to school and it means a lot to them."

Then again, Inez likes to teach because it's just plain fun. "I like the material," she says. "I really like my equations. I like telling freshmen what the physics is, what the biology is. I like explaining why it's important. It's the same way when you see a great movie or read a great book. You want to tell everyone about it."

And if you take one of Inez's courses, you never know who might drop by. She often invites scientist friends to visit and tell the students stories about their discoveries and travels.

Inez worked on projects with Piers Sellers when they were both NASA scientists in the 1980s and early 1990s. Then Piers followed a lifelong dream and became an astronaut in 1996. He flew on the space shuttle *Atlantis* and made several walks in

space. When Piers was passing through Berkeley, he came to Inez's class and thrilled the students with tales of his experiences in space. He was happy to do it for his old friend.

"She's a very good scientist," Piers says. "But she's also the model friend. She's considerate and humorous—really good company. People love to be around her and support her, and are happy for her success."

~ Going to Bat for the Environment

As a teacher, Inez thinks the same way she thinks as a climate researcher. She looks at the big picture and tries to find answers to questions that interest her. In teaching students, Inez wondered, "What makes a well-educated citizen?" Whenever she answered the question, her definition included knowledge of the natural environment.

As a result, Inez's newest passion is environmental education. She believes that understanding the environment is a crucial part of being a well-educated person. It's as important as knowing history and math.

For the last couple of years, Inez has been one of the leaders of a team trying to start a new environmental education program at UC Berkeley. They hope to make the study of the environment part of *every* student's education. "We are so dependent on Earth's natural systems," Inez explains. "We need to understand how the system supports us." As part of this initiative, she would like to help teams of the top people in various fields at UC Berkeley work together to solve important environmental problems.

> In the freshman seminar every week, Inez has lunch with the class. They talk about the atmosphere over sandwiches.

Inez understands the way Earth's natural systems work. She also understands what can happen when the actions of people throw these systems out of balance. It leads to the depletion of resources, the killing of plant and animal life, or even the changing of climate patterns. "I have tremendous love and respect for the planet—how the whole system comes together," Inez says. "That's why this is important. A public that understands the environment will understand what sustains them. People need to know the long-term implications of what they do each day. They need to know how it affects the Earth system that allows life to exist."

~ A Woman of Honors

Inez has been honored several times in her career for pioneering research in the field of global climate change. In 2001 she was elected to the National Academy of Sciences. "My parents were proud in their quiet way," she says. "They were too old to travel to my induction into the academy. I wore my mother's dress to the ceremony, and she said she wished that her mother was alive to see it."

As part of her election to the National Academy of Sciences, Inez received a diploma *(above)*. At left Inez poses with two other women who were also inducted: plant biologist Pat Zambryski *(middle)* and biomechanist Mimi Koehl.

In December of 2004 Inez received the Revelle Medal from the American Geophysical Union, the world's largest society of Earth scientists. The Revelle Medal recognizes outstanding accomplishments and contributions in the Earth sciences. It was named for Roger Revelle, the oceanographer who had led the charge in warning about the possible dangers of global warming.

Inez became only the second senior woman medalist. She was introduced at the ceremony with these words: "Inez is a scientific pioneer. She has often made important discoveries on topics before most have even recognized them as important areas of pursuit. Along the way, she has laid the groundwork for the new area of biogeoscience. She has also developed many of the key modeling and numerical analysis techniques in use today."

In 2004 John Orcutt, president of the American Geophysical Union, presented Inez with the Revelle Medal *(above)*. Inez *(front row, center)* often attends international conferences such as this one at the Pontifical Academy of Sciences in Vatican City in 1998.

Inez has a special respect and affection for Roger Revelle. He was the chair of the first science committee she served on. After working with her, Revelle himself described Inez as a "remarkable" woman. When she received the award, Inez talked about Revelle's impact as a scientist. He studied global warming long before many other scientists recognized its importance. But she also talked about Revelle's little-known work in Pakistan when John F. Kennedy was president. Revelle helped set up irrigation systems so the poor farmers of that nation could grow more food. She said this work was an example of "geoscience for peace," which she finds inspiring.

~ A Friend Indeed

More important to Inez than awards is the circle of companions she has had since her student days at MIT 30 years ago. They talk to each other about work and are there for each other as friends.

Inez is the organizer of the group, the one who plans outings that draw these busy scientists together from all parts of the country. When a member of the group turns 60, Inez plans a party for the occasion.

A birthday celebration in Venice, Italy, marked the 60th birthday of Eugenia Kalnay. "It was an example of her generosity and her organization skills," says Eugenia, a meteorology professor. "She brought everyone together and we had dinner. It was lovely."

Ed Sarachik remembers his 60th birthday party, at a winery in California's Napa Valley. "She just does these kinds of things for people," he says. "She also organized a dinner last December at a really fine restaurant in San Francisco." That was to celebrate the birthdays of Mark Cane and Jagadish Shukla.

Mark remembers that Inez was the Saturday events organizer of the group even back in graduate school days at MIT. "She is the one who makes sure we all keep in contact," he says. "She's very caring and loyal."

~ A Birthday Challenge

Inez has always treasured her friends and colleagues. Above, she is flanked by friends from her MIT days, Antonio Moura *(left)* and Jagadish Shukla. At top, Inez *(fourth from left)* relaxes with Australian scientists.

Inez celebrated her 56th birthday in 2005. She is already planning her 60th. On that landmark day, she plans to swim across San Francisco Bay. She will start at Alcatraz Island and end up at Fisherman's Wharf in San Francisco. Right now, she's scoping out pools to practice in and thinking about buying a wetsuit. You need one to keep you warm in the cold, rough waters of the bay. "I haven't gotten wet yet," she says, "but I have almost five years to train."

"You go in a group with a club," she explains. "I have to find a club for old, sedentary people," she jokes. Is she worried about keeping up? Not really. It's a swim of just a couple of hours, says the woman who swam regularly in the warm waters of the South China Sea as a girl.

Inez likes the idea because it's a challenge—and because it will be fun. So far, none of her friends have agreed to go along. "My husband said he'll follow me in a kayak," Inez says, laughing at the thought. "He just doesn't float very well." Old friend Mark Cane—the person who first suggested she work at GISS almost 30 years ago—has said he might swim with her. If, that is, she'll do it in slightly warmer waters. Florida, maybe?

Will Inez go through with her birthday swimming adventure? Who knows? But when Inez Fung sets her mind to something, it's wise not to bet against her.

Timeline of Inez Fung's Life

1949   Inez Fung is born on April 11 in Hong Kong.

1967   Inez graduates from Kings College, a college preparatory school for math and science students. Following political riots in Hong Kong, Inez's parents send their children to schools in North America.

1971   Inez earns a bachelor's degree in applied mathematics from the Massachusetts Institute of Technology (MIT).

1976   Jim Bishop, a fellow graduate student at MIT, and Inez wed in Vancouver, Canada, where their parents live.

1977   Inez graduates from MIT with a doctor of science degree in meteorology. She was then only the second woman to do so in that MIT department. Her thesis wins the Rossby Award for the best meteorology thesis of the year. She begins postdoctoral work at NASA, at the Goddard Space Flight Center (GSFC) in Greenbelt, Maryland.

1979   Inez is transferred to the Goddard Institute for Space Studies (GISS) in New York City.

1981   Inez creates her first three-dimensional global carbon model.

1986   Inez becomes an American citizen.

1987   For her work with two other NASA scientists on carbon cycling, Inez wins NASA's Exceptional Scientific Achievement Medal.

1990   Working with two other scientists, Inez makes a landmark discovery about "disappearing" carbon dioxide in the atmosphere.

1991   Inez publishes a three-dimensional global model of methane gas.

1993   Jim and Inez move to Vancouver Island in Canada to teach at the University of Victoria.

1995   Inez teams up with postdoc Ina Tegen to develop a dust model that links dust in the atmosphere to climate change.

1997   Inez becomes the first director of Berkeley's Center for Atmospheric Studies at the University of California.

2001   Inez is elected to the National Academy of Sciences.

2004   For her outstanding contributions to Earth science, the American Geophysical Union awards Inez the Revelle Medal.

2005   Inez and Jim live in Berkeley, where he is an oceanographer and she continues to teach. In 2009 Inez plans to swim across San Francisco Bay in celebration of her 60th birthday.

ABOUT THE AUTHOR

Renee Skelton has a great interest in science and enjoys writing about it. She has written several books for children on subjects such as the natural environment and the lives of noteworthy scientists. She was an editor and writer for Sesame Workshop's science magazine *3-2-1 Contact*. Renee has also contributed science articles to many children's magazines such as National Geographic's *National Geographic Kids*, Time Inc.'s *Time for Kids*, and Scholastic Inc.'s *SuperScience*. Renee lives in New Jersey.

GLOSSARY

This book is about a scientist who studies conditions in the atmosphere that cause climate change. The atmosphere is the layer of gases that surround a planet. The word *atmosphere* comes from the Greek *atmos,* meaning "vapor," and the Latin *sphaera,* meaning "sphere." Many scientific terms have Greek or Latin origins.

Here are some other science words you will find in this book. For more information about them consult your dictionary.

biogeochemical cycles: cycles that support life on Earth by moving important chemicals among the land, air, oceans, and living things

biosphere: the living part of Earth's environment, including plants and animals

carbon cycle: the constant movement of carbon, in its many forms, between the nonliving environment (oceans, soil, atmosphere) and living things

carbon dioxide: a gas that occurs naturally in Earth's atmosphere and as the result of human activities such as the burning of fossil fuels. It is a main cause of global warming.

climate: the average weather conditions of an area or region occurring over a period of years. Climate conditions include temperature, precipitation (rainfall or snowfall), and wind speed.

climate model: a computer program that simulates the conditions that determine climate

cloud: a collection of visible particles suspended in air. In Earth's atmosphere, clouds are composed mainly of tiny water droplets or ice crystals.

cycle: a series of events or actions that occur regularly and usually lead back to a starting point. From the Greek *kyklos,* meaning "circle."

Earth systems model: a computer program that simulates the atmosphere, oceans, and one or more of the biogeochemical cycles that support life on Earth

ecology: the study of the relationships among plants, animals, and other living things and their environment. From the Greek *oik,* meaning "environment."

ecosystem: a community of living organisms interacting with one another and with their nonliving environment

fluid dynamics: a branch of science that deals with the movement of liquids (such as ocean waters) and gases (such as those in the atmosphere). From the Latin *fluere*, meaning "to flow" and the Greek *dynamis*, meaning "power."

fossil fuel: a fuel such as coal, petroleum, or natural gas, formed from living organisms that died millions of years ago. Fossil fuels contain carbon. Carbon dioxide gas is released as a by-product of burning fossil fuels.

global warming: a rise in the temperature of Earth's surface and lower atmosphere caused by an increase in the amount of carbon dioxide and other greenhouse gases in the atmosphere

greenhouse effect: the process that keeps Earth's lower atmosphere warm enough to make life on the planet possible. Like the glass in a greenhouse, carbon dioxide and other gases in the atmosphere trap infrared radiation and send it back toward Earth where it warms Earth's surface and the lower layers of the atmosphere.

ice age: one of several periods of Earth's history when the average global temperature decreased and glaciers covered large areas of land

meteorology: the study of the atmosphere and atmospheric conditions, especially as they relate to weather

ocean current: a rapid stream of water moving in a certain direction within the ocean

photosynthesis: the process in which cells in green plants combine carbon dioxide from the air and water and nutrients from soil, in the presence of sunlight, to make carbohydrates, such as glucose, for plant growth. Oxygen is released as a by-product of photosynthesis.

respiration: the process in which animals and plants take up oxygen from the environment and convert it to provide energy for their living cells. In plants, the carbohydrates made during photosynthesis, such as glucose, combine with oxygen to release energy, water, and carbon dioxide.

sink: any body or process that removes a particular chemical from the atmosphere. The ocean is a sink for carbon dioxide.

transpiration: the process in which water vapor leaves plants through tiny holes in their leaves

typhoon: a violent tropical cyclone in the western Pacific Ocean with winds of at least 74 miles per hour. In North America typhoons are called hurricanes.

weather: the state of the atmosphere at a certain place and time in terms of variables such as temperature, precipitation, wind, and cloudiness

Conversion Chart

When you know:	Multiply by:	To convert to:
Inches	2.54	Centimeters
Miles	1.61	Kilometers
Centimeters	0.39	Inches
Kilometers	0.62	Miles

When you know:	Use this formula:	To convert to:
°F	$(5/9) \times (°F - 32)$	°C
°C	$(1.8 \times °C) + 32$	°F

FURTHER RESOURCES

Women's Adventures in Science on the Web

Now that you've met Inez Fung and learned all about her work, are you wondering what it would be like to be a climate scientist? How about an astronomer, a wildlife biologist, or a robot designer? It's easy to find out. Just visit the *Women's Adventures in Science* Web site at **www.iWASwondering.org**. There you can live your own exciting science adventure. Play games, enjoy comics, and practice being a scientist. While you're having fun, you'll also get to meet amazing women scientists who are changing our world.

BOOKS

Parks, Peggy J. *Global Warming*. Lucent Library of Science and Technology. San Diego, CA: Lucent Books, 2003. This book offers a historical perspective of global warming, from the Industrial Age to the present. It also presents the thoughts of scientists who are alarmed about the issue—and those who are not.

Pringle, Laurence P. *Global Warming: The Threat of Earth's Changing Climate*. New York: SeaStar Books, 2003. This book includes information on nuclear power, the carbon cycle, coastal lands lost to rising sea levels, the effect of El Niño, and the dangers of aerosols. It also covers the steps that have been taken to slow the warming of Earth's atmosphere. Many photos and large colorful maps and diagrams are also included.

WEB SITES

EPA: http://yosemite.epa.gov/oar/globalwarming.nsf/content/index.html
The U.S. Government's Environmental Protection Agency site is a good place to go to have your questions answered about global warming, including how serious the situation is now and what you can do to lessen your contribution to the problem.

NASA's Earth Observatory:
http://earthobservatory.nasa.gov/Library/GlobalWarming/
Learn about the many possible effects of global warming and how climate models help scientists determine future climate. It's also worth visiting the Earth Observatory's home page at http://earthobservatory.nasa.gov/ where you can view satellite images and learn lots more about our home planet.

NOAA: www.cmdl.noaa.gov/infodata/faq.php
The National Oceanic and Atmospheric Administration has a climate monitoring and diagnostic laboratory that features frequently asked questions. Find out how scientists predict future climate, how global warming can affect us, and what causes climate change.

SELECTED BIBLIOGRAPHY

In addition to interviews with Inez Fung, her family, and her friends, the author also did extensive reading and research to write this book. Here are some of the sources she consulted.

McKnight, Tom L., and Darrel Hess. *Physical Geography: A Landscape Appreciation.* 8th ed. Upper Saddle River, NJ: Pearson Prentice Hall, 2005.

Miller, G. Tyler. *Living in the Environment.* 13th ed. Pacific Grove, CA: Brooks Cole, 2004.

Moran, Joseph M., and Michael D. Morgan. *Meteorology: The Atmosphere and the Science of Weather.* 5th ed. Upper Saddle River, NJ: Prentice Hall, 1997.

National Research Council, Division of Earth and Life Sciences, Committee on the Science of Climate Change. *Climate Change: An Analysis of Some Key Questions.* Washington, DC: National Academy Press, 2001.

Tarbuck, Edward J., and Frederick K. Lutgens. *Foundations of Earth Science.* 10th ed. Upper Saddle River, NJ: Prentice Hall, 2003.

Weart, Spencer R. *The Discovery of Global Warming.* Cambridge: Harvard University Press, 2003.

INDEX

LIBRARY ADVISORY BOARD

STUDENT ADVISORY BOARD

Illustration Credits:

Except as noted, all photos courtesy Inez Fung

Cover photo of Ms. Fung, Margaret Gennaro; **2** © 1998 Corbis; **3** © 1999 PhotoDisc; **5** (*t*) Rob Wood, Matthew Frey, Wood Ronsaville Harlin, Inc.; (*b*) Courtesy BC Greenhouse Builders LTD; **6** (*t*) Morton Elrod, Glacier National Park Archives; (*b*) Lisa McKeon, USGS; **9** (*t*) AP/Wide World Photos; **11** © Hong Kong Observatory; **12** inset map of Hong Kong, Will Mason; **18** (*t*) © Bettmann/Corbis; **21** © 1999 PhotoDisc; **22** Courtesy Archives, Utica College; **24** Courtesy MIT Museum; **25** Donna Coveney/MIT; **26** Courtesy MIT Museum; **32** (*l*) NASA/GSFC and Orbimage; (*r*) © 1994 Digital Stock; **33** (*t*) Courtesy MIT Museum; (*b*) U.S. Army Ordnance Museum; **43** © 1992 PhotoDisc; **45** (*t*) NASA; (*b*) Terry J. Adams, NPS; **48** © Goodshoot; **49** © 1997 PhotoDisc; **50** (*t*) NASA; (*b*) NOAA; **53** (*t*) © 1998 Digital Stock; (*b*) NASA; **54** (*t*) NASA; (*b*) NPS; **59** Svante Arrhenius Archive, The Center for History of Science, The Royal Swedish Academy of Sciences; **60** (*t*) AIP Emilio Segrè Visual Archives; (*b*) NOAA; **61** (*t*) NOAA; (*b*) Courtesy National Academy of Sciences; **65** (*t*) NASA/GSFC and Orbimage; (*b*) strawberry plant, © Brand X Pictures; compost, Courtesy Texas Cooperative Extension; cow, © 1998 Corbis; coastline, NOAA; coal, © 1999 Corbis; **66** (*t*) Copyright © 1988 by the New York Times Company. Reprinted by Permission; (*b*) Jose R. Lopez/The New York Times; **68** © 1999 Corbis; **74** (*t*) © 2000 PhotoDisc; (*b*) USDA; **79** (*t*) © 1998 Digital Stock; (*b*) © 1995 PhotoDisc; **83** NOAA; **84** Todd Sowers, LDEO; **85** NSF Office of Polar Programs and USGS; **90** James H. Pickerell; **93** (*b*) Courtesy University of California, Berkeley; **95** (*b*) Jim Bishop

Maps: © 2004 Map Resources

Illustrations: Max-Karl Winkler

The border image used throughout the book is of clouds, © 1998 Digital Stock.

JHP Executive Editor: Stephen Mautner

Series Managing Editor: Terrell D. Smith

Designer: Francesca Moghari

Illustration research: Christine Hauser

Special contributors: Roberta Conlan, Meredith DeSousa, Allan Fallow, Sally Groom, Mary Kalamaras, Dorothy Lewis, April Luehmann, John Quackenbush, Anita Schwartz, Amanda Staudt

Graphic design assistance: Michael Dudzik and Anne Rogers